19th Century
AMERICAN CARRIAGES
*Their Manufacture, Decoration
and Use*

19th Century
AMERICAN
CARRIAGES
*Their Manufacture, Decoration
and Use*

PETRI-PEES-N.Y.

THE MUSEUMS AT STONY BROOK

This publication was made possible by a grant from the National Endowment for the Humanities, a Federal agency.

The Museums at Stony Brook

Editorial assistance by Alison Kilgour
Design by Melanie Roher Design, Inc./The Whelan Design Group
Photography by Lynton Gardiner, unless otherwise noted
Black-and-white lab services by Kenneth S. Taranto, New Jersey;
Color lab services by Cosmic Sound-Delight, New York, unless otherwise noted
Word Processing by Absolute Priority Processing, Inc.
Typography by Centennial Graphics, Inc.
Printed by Eastern Press, Inc.

Published by The Museums at Stony Brook
1208 Route 25A
Stony Brook, New York 11790

Library of Congress Cataloging-in-Publication Data
19th-century American carriages.
  Includes bibliographies.
  1. Carriages and carts—United States—History—19th century. 2. Carriage industry—United States—History—19th century. I. Museums at Stony Brook. II. Title: Nineteenth-century American carriages.
TS2010.A13   1987   688.6   87-12282
ISBN 0-943924-10-3

Cover: "The Centennial—Carriage Exhibit"
from sketches by Theo. R. Davis
Harper's Weekly, January 20, 1877

CONTENTS

FOREWORD

The Carriage Era lasted for a relatively brief period, from the late seventeenth century until the automobile essentially replaced the carriage in the first decades of the twentieth century. In the seventeenth and eighteenth centuries, carriages were extremely expensive to own and maintain and consequently were a scarce commodity. Because roads were poor and carriage suspension systems rather primitive, riding in a carriage was not very comfortable. In the course of the nineteenth century industrialization had a major effect on the production, design and use of horse-drawn vehicles. This was particularly true in the United States, where industrialization and the ingenuity of individual carriage makers and designers made possible the production of a wide range of types of vehicles, some based on European styles but many developed in the United States.

The Museums carriage collection is widely representative, reflecting the diversity of types of vehicles that were used during the Carriage Era primarily, although not exclusively, in the United States. This is due to the deliberate shaping of the collection by Ward Melville in the 1940s and 1950s. He was ably assisted in the development of the collection by Richard McCandless Gipson. The companion volume to this publication, *The Carriage Collection*, provides a brief history of the collection's formation as well as detailed descriptions of 80 vehicles, which were selected to be reprentative of the entire collection of over 250 vehicles.

Melville's and Gipson's foresight in creating a collection reflective of historical patterns of design, production, ownership and use of horse-drawn vehicles provides us today with a primary artifactual resource for the study of a major aspect of America's transportation history. The collection provoked the initial questions that led to this collection of essays. It also provides visual documentation and information that, when combined with information gleaned through traditional research methods, provides a richly textured picture of carriage history in nineteenth-century America.

The history of horse-drawn transportation is just beginning to be explored. Most published sources on the subject date from the last decades of the nineteenth century or the first decades of the twentieth. (See "Further Readings" in *The Carriage Collection* for a listing of major publications on carriages and carriage history.) This book shares new research on the subject, based on the study of primary sources such as tax records and probate inventories; the subject is large and the research resources vast and often scattered. These essays have perforce been sharply focused: This volume does not attempt to tell the entire history of horse-drawn transportation in the United States. The major focus of the essays is on the nineteenth century, and on the northeastern section of the United States (although the importance of the Midwest in the American carriage industry is also addressed). Two of the four essays focus even more particularly on New York and on Long Island within that state.

In the interest of historical accuracy all original material is printed as it was written. Sic is not used. Illegible, questionable or contextually suggested information is enclosed in brackets. Illustrations, unless otherwise noted, are of objects in The Museums Carriage Collection or in The Museums Carriage Reference Library.

Encouragement, support and assistance have been provided by many since the beginning of this project in 1981. The National Endowment for the Humanities, a

Federal agency, provided a major grant for the planning, research and production of a two-volume publication: the first volume a catalog of The Museums carriage collection and the second a book of essays on aspects of the history of carriages. By participating in several planning conferences that were partially funded by both the National Endowment for the Humanities and by the New York State Council on the Arts, a number of scholars helped to shape both the questions to be addressed and the research resources and methodologies for seeking answers to these questions. The Museums acknowledges with warm appreciation the advice and encouragement of these scholars:

David Hackett Fischer, who assisted the project through active involvement in all of the planning conferences, and Michael J. Ettema, Larry Lankton, Roger Parks, Richard E. Slavin, III, Maris Vinovskis, Richmond Williams and Michael Zuckerman, each of whom participated in at least one of the conferences.

Encouragement and assistance for the project were given by several members of the History Department of the State University of New York at Stony Brook, especially Nancy Tomes, Ned Landsman, Eric Lampard and Karl Bottigheimer.

The advice of The Museums Carriage Advisory Committee was most helpful in developing both volumes of this publication. The committee was chaired by James Blackwell during the preparation of this book; his untimely death in February, 1986, deprived The Museums of his invaluable leadership and generous assistance in matters relating to The Museums carriage collection.

Many institutions and people within those institutions were of particular help in the research for this book, including the following:

Albany Institute of History and Art, Albany, New York; Don Berkebile, Mercersburg, Pennsylvania; Clinton County Historical Society, Plattsburgh, New York (Helen W. Allan); Discovery Hall Museum, South Bend, Indiana (Julie K. Bennett and Marsha A. Mullin); East Hampton Free Library, Long Island Room (Dorothy King); Federal Records Center, Bayonne, New Jersey (Anthony Fantozzi); Henry Ford Museum and Greenfield Village, Dearborn, Michigan (Randy Mason); Huntington Historical Society, Huntington, New York (Irene Sniffin); Mary Littauer, Syosset, New York; The Metropolitan Museum of Art, New York, New York; Minnesota Historical Society, St. Paul, Minnesota (John Wickre); Nassau County Museum, East Meadow, New York (Gary Hammond); National Archives, Washington, D. C. (William Scherman); National Museum of American History, Smithsonian Institution, Washington, D. C. (William Worthington); New Hampshire Historical Society, Concord, New Hampshire (William Copely); The New-York Historical Society, New York, New York; New York Public Library, Division of Special Collections and Division of Rare Books and Manuscripts; New York State Archives, Albany, New York (James Folts); New York State Department of Agriculture and Marketing, Albany, New York (Donald Keating); Ontario County Historical Society, Canandaigua, New York (Virginia L. Bartos); The Queens Borough Public Library, Long Island Division, Jamaica, New York (Robert Friedrich); Thomas Ryder, Salem, New Jersey; St. Joseph's College, Long Island Room, Patchogue, New York (Sister Joan Ryan); The Smithtown Library, The Long Island Room, Smithtown, New York (Vera Toman); State University of New York at Stony Brook, Special Collections, Stony Brook, New York (Lee Hiltzik); Staten Island Historical Society, Staten Island, New York (Stephen Barto); Suffolk County Historical Society, Riverhead, New York (Joanne Brooks); State Historical Society of Wisconsin (Jean Weber). The Museums is grateful for their help in providing a wealth of information about horse-drawn vehicles in nineteenth-century America for the preparation of this book.

FROM CARRIAGE SHOP TO CARRIAGE FACTORY:

The Effect of Industrialization on the Process of Manufacturing Carriages

Joanne Abel Goldman

Sellier-Carrossier
1751-1765
Selles Illustrator,
A. B. Carrochi F.
From Encyclopedie
by Denis Diderot
Paris, France

A French carriage-making shop, such as is seen in this illustration, was undoubtedly larger and more sophisticated than any carriage-making shop in America in the mid-eighteenth century. Many European carriage makers were artisans who also practiced decorative arts and interior decoration.

The second half of the nineteenth century was a time of transition in the carriage industry:[1] The themes of growth and diversity characterize the changes that revolutionized both the process and products of carriage manufacture. Between 1850 and 1900 the volume of production doubled and trebled, and this prosperity was due in large part to the adoption of the factory system of production.

In a factory, production is centralized. Power, machinery and labor are all assembled under one roof so that the manufacture of a particular product can be carried out at one locale. Two distinct types of factories developed during the nineteenth century. One concentrated on the production of a particular carriage part, and can therefore be called a specialty factory. The parts produced by this type of factory were sold to companies and shops where the pieces were assembled, thus actually building the vehicle. One observer remarked of this work, "They buy a body here, the wheels there, the axles in New Jersey, the springs in Connecticut; they clap them together by machine power, cover them with paint bought ready mixed, and finish them with decalcomanie coat-of-arms, which were painted in Germany".[2]

The other type of carriage factory was the vertically integrated factory. In this factory all of the work processes required to construct a carriage were performed. Although both types of factory came to dominate production in the post–Civil War years, they did not prevent the traditional carriage shop from continuing to command a considerable percentage of the market; the traditional method of production continued throughout the nineteenth century. This essay, however, will focus on the factory system of production.

THE SHOP TRADITION

Before industrialization and the subsequent development of the factory, it was the carriage shop that dominated production; as the nineteenth century opened, carriage manufacture took place in relatively small owner-operated shops. Production was carried out with the use of general-purpose machinery and hand tools and power was supplied by animal, water or manpower. The actual work involved in the production of a carriage was carried out by a master carriage maker, who was usually assisted by one or more apprentices.

These craftsmen were trained in the traditional apprentice system, in which young men were contracted to master artisans for a number of years to learn a trade. In the carriage trade this education included learning such skills as blacksmithing, body-making, finishing, machining, painting, stitching, striping, trimming, varnishing and wheelwrighting. Although occasionally a carriage maker would depend upon a specially-trained artisan to carry out a particular work process, or may have had apprentices or journeymen with specialized skills working for him, for the most part a carriage maker was adept at all of the skills needed to build a carriage.

As the nineteenth century progressed,

Basket wagon
c. 1810
Maker unknown,
probably United States
Gift of William Jarvis, Jr.
in memory of
Lucretia Jarvis, 1957

This vehicle, in which General Lafayette traveled in New Hampshire in 1825, includes a variety of materials–woven willow for the body, silver-plated metal handles and dash rail, and several textiles: velvet, silk and linsey-woolsey, some of which were probably imported from Europe. While the style of this wagon is not as sophisticated as that of comparable European vehicles, workmanship reflecting a variety of carriage-making skills–blacksmithing, body-making, wheelwrighting, trimming (applying the interior and exterior fabrics and leather, similar to upholstering)–is excellent.

an increase in demand for horse-drawn vehicles stimulated productivity and created a need for a level of production that was unobtainable in the small owner-operated shop. The factory and the factory system of production were able to meet the increased demand and thus came to dominate manufacturing.

THE CARRIAGE FACTORY AND THE FACTORY SYSTEM OF PRODUCTION

The production of carriages in a late-nineteenth-century carriage factory was a much more capital-intensive endeavor than the process of traditional carriage making had been. To a large extent, much of the high cost of production was due to the added overhead of machinery. The most basic as well as the most necessary of machines was a power plant that could supply a safe, reliable and centralized source of power.

Although in the past power had been supplied by animals or water, the former were neither reliable nor consistent and the latter required locating factories at particular and perhaps inconvenient sites. By the 1870s the steam engine was generally adopted by the carriage industry. Its popularity was such that in 1880 the U.S. Census of Manufacture reported it to be the most widely-used power source in the United States. However, by the turn of the century, its supremacy was rivaled by plant-generated electricity. One of the advantages of electric power was that it created less vibration and noise and was more easily managed, safer and more reliable than steam engines.[3]

Once a source of energy became available, machinery could be used to assist many of the work processes. This is evidenced by the relatively early yet widespread appearance of machinery in the factory. While in 1820 the average carriage shop was equipped with only lathes, forges and the hand tools of the blacksmith and wheelwright, a late-nineteenth-century carriage factory included a much more extensive array of equipment.[4] The inventory of one such factory in 1870 included fifteen circular saws, two wood lathes, six iron lathes, six tenoning and boring machines, three planers, one box setting, six polishing stones, six top hangers, two spring machines and four drills, all powered by a steam engine. The company

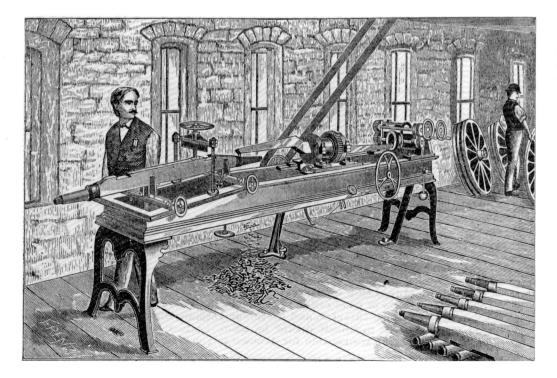

Setting Skeins by Machine
c. 1880
From catalog,
Milburn Wagon Company
Toledo, Ohio

reported $200,000 in capital costs[5]–a significant increase over the $125 that its proprietor had initially invested into his business 57 years earlier.[6]

The use of machinery permitted many changes in production, the most significant of which was that it both made possible and depended upon the standardization of parts. Although the concept of standardized parts dated back to the arms industry of the late eighteenth century, it was not applied to carriage production until the mid-nineteenth century. The traditional reliance upon the hand and eye was contrary to the standardization of parts, for it gave each piece unique characteristics; once machines were used for production, and standards for gauging, however, duplication was possible to a degree unattainable with non-mechanized fabrication. Furthermore, the standardization of parts permitted even further mechanization in the process of carriage production. For example, the wheelwright's machine, patented in 1858, relied upon the standardization of the wheel. Once wheel sizes were limited to a predetermined variety, this machine could perform the double duty of tenoning and hub boring. Standardized parts also facilitated repair because they no longer had to be individually fitted; furthermore, they could be relied upon to fit into other standardized parts, thus enabling as well as encouraging mass production.

By the 1830s specialized tools for carriage manufacture began to appear in the United States patent records. These tools allowed the development of techniques uniquely suited to American materials and conditions. Similarly, such machinery allowed the exploitation of new materials for carriage manufacture. Of particular importance was the hickory lathe; it facilitated the construction of hickory wheels, which were far superior to the heavier European wheels, made of a variety of woods including oak. Since hickory was both stronger and lighter than oak, it significantly lightened the vehicle, thus making the carriage more manageable on the often rough American roads.

Mechanization encouraged the growth and concentration of the industry into large factories, and a cyclical economic relationship soon developed. A growing demand was satisfied by an increase of production which was made possible by the reliance of manufacturers on the capital-intensive factory method of production. These factories in turn could only be supported by high-volume production—which, incidentally, the market was able to absorb.

A variety of types of the specialty factory soon developed, ranging from those which made entire carriage bodies to those which manufactured carriage trimming material or varnish. Any item needed for carriage production could be purchased from one of these establishments; they produced, among other components, wheels, axles, boxing machines, carriage couplings, whip sockets, name plates, hubs, trimming and finishing hardware, gears, coach laces and coach lamps. This method of production was applauded by Ezra Stratton, a prominent carriage builder and trade journal publisher, in 1878:

Among the leading causes of this improvement [of the industry], we must, first of all, mention the

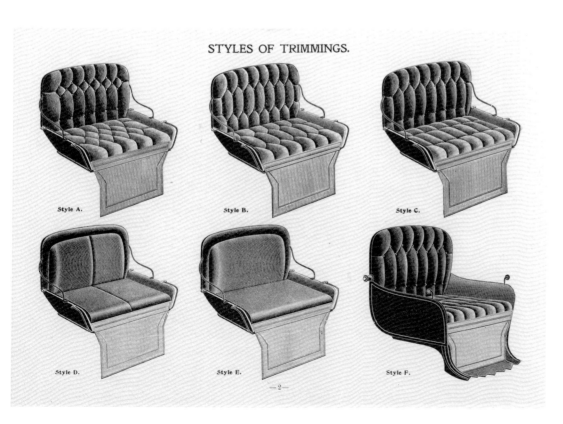

STYLES OF TRIMMINGS.

Style A.

Style B.

Style C.

Style D.

Style E.

Style F.

—2—

Styles of Trimming
c. 1890
From catalog,
Muncie Jobbing and
Manufacturing Company
Muncie, Indiana

division of labor which now prevails in the manufacture of the various component parts of a carriage. Axles, springs, clips, bolts, and all other iron parts, which every carriage builder was formerly compelled to manufacture for himself, with such means as he had at his command, are now produced in large quantities in special establishments, with a remarkable degree of perfection, and at prices much below the former ones.[7]

In addition, a great deal of production was carried out in the vertically integrated carriage factory, an example of which was New York City's Healey, Williams and Co. carriage factory. Its physical description is rather telling. In the basement were the steam engines, the smith shop and the machine shop. The ground floor was devoted to the office, stockroom and carriage ware-room or showroom. The second floor was divided into a trimming and a finishing department. The third floor was occupied by the body, gear and wheel makers. The fourth floor was devoted to the seasoning and storing of timber and most of the woodworking. The fifth floor was where the paints were prepared and the gears were painted. The sixth floor provided space for painting bodies which were then dried on the partially covered roof.[8]

Inside the factory the labor force was divided into individual departments in accordance with particular work processes. By dividing and subdividing work in such a manner that each man had but a single part to perform, a level of efficiency that was unattainable by traditional methods could be achieved,[9] facilitating an increase in both production and profits. The product, i.e., the carriage, was made of many different materials and required multiple specialized skills to fabricate, and was therefore a near-ideal object to be produced by the subdivision of labor.

By the close of the nineteenth century the methods of carriage manufacture were both numerous and varied; the specialty factory, the vertically integrated factory and the small and often rather traditionally run carriage shop made up an industry that continued to grow, prosper and diversify into the twentieth century.

RESPONSES TO INDUSTRIALIZATION

After the Civil War, industrialism and mechanization were pervasive, and people within the carriage industry were noting changes in the traditional work processes. In 1872 C. P. Kimball, a particularly distinguished carriage builder, remarked that "New labor-saving machines of all kinds have been introduced, and every aid science and ingenuity can invent [is] brought into requisition."[10]

People responded to these changes in a variety of ways. Some welcomed them with eagerness while others reluctantly accepted the inevitable with great trepidation. By 1870 the changes that were taking place were the subject of much debate. While some praised machines for enabling a better carriage to be built, others argued that the quality of the work that was done by machines was not of the same grade as that produced by the craftsman's trained hand. Respect and acclaim for the talents of pre-industrial craftsmen were prompted by the fear that industrialization would eventually eliminate the need for their skills. Naturally, the artisans themselves worried about this. One contemporary expressed this mood rather explicitly:

It is fact, though perhaps not generally known that, now and then and here and there, we may still find a carriage-painter out of whom the wholesale factories,—those formidable and relentless enemies of skilled labor,—have not utterly crushed the last spark of ambition but in whom there still remains a lingering desire to rise in his chosen calling.[11]

The concern expressed in this passage was not without due cause, for machines did automate a great many skills and did eliminate work for the artisan. He was, at least to some degree, replaced by mechanics and unskilled laborers, although in departments of some factories the artisan still held a place of prominence, and was generally appointed foreman in most establishments.

In addition to the effect that mechanization had upon the traditional skills of carriage making, other consequences of industrialization also caused anxiety. Leaders of the industry were concerned that the use of machines would increase the volume of production, inevitably leading to overproduction, which would cause a decrease in prices and the subsequent loss of capital and jobs.[12] These fears were expressed by Mr. J. L. H. Mosier, who argued in 1876 that even though the employment of machinery might be extremely advantageous because it increased productivity, its effects would inevitably hurt labor. He pointed out that the drilling machine was 33% more productive than hand power. The bolt cutting machine "cuts and nuts" 60 bolts in one hour as opposed to the 30 bolts in ten hours that could be produced by hand. The use of machines could–and Mr. Mosier feared would–cut the labor force considerably. Not only would fewer workers be needed, but unskilled "boys" rather than trained craftsmen could fill many positions, and for smaller wages. For example, the bolt-cutting machine would yield the desired increase in productivity whether it was run by a boy or a man. However, because an unskilled worker commanded a much lower salary than a "smart man"–$6 per week as opposed to $10 per week–the adult would probably lose his job to the competition of child labor.[13]

Although these concerns about changes in the character of the industry caused some to resist "progress," others welcomed industrialization. They looked forward to the end of grueling work processes and applauded the innovations that industrialists were incorporating into their factories.

In those earlier days the sweat of man's brow moistened every portion of the work. Now a few drops of oil on a noiseless machine are all the lubrication required. Then a coachmaker must know how to use a multitude of tools–must be able to drive the joiner, to draw the shave and to push the triangular chisel. Now he can get along if he but know how to adjust a lever or to ship a belt. Revolution has taken place in every department of the works . . . Every department [of the factory] tells the same story. Machinery has come to man's assistance, increasing his powers and lessening his toil.[14]

Despite the mixed response during the period of transition, the pace of progress continued since the promise of profit, together with a tremendous increase in production, was a strong motivation for invention,

innovation and industrialization. One observer chose a particularly interesting way of bridging the apparently conflicting responses to the changing times. He suggested that rather than mechanization replacing the skill of the craftsman, it depended upon his expertise:

An educated man finds out the value of machinery and desires to use and improve it. Instead of fearing its rivalry he welcomes it; he remembers that all tools, even the saw and the hammmer, are machines, and that the hand, the human hand that guides those tools is but a perfect machine obeying the guidance of the brain more quickly and in a more varied manner than any man-made machine. The American workman therefore uses machines more and more.[15]

GROWTH OF THE INDUSTRY

By the end of the nineteenth century the nature of the carriage had changed considerably. Once an expensive and luxurious work of art used almost solely by the gentry class, by the 1870s it was both affordable and widely available in its mass-produced form. Factory production, whether within the walls of a vertically integrated carriage factory or in an assemblage of parts made in specialized carriage parts manufactories, brought about the "democratization" of the carriage. By increasing the rate of production as well as the efficiency of the process, the price of the vehicle could be reduced resulting in increased affordability and popularity. Trade journals praised the American system that allowed the production of the "cheapest and best light vehicles for the money that are produced in the world" and boasted that "every man among us who can afford to keep a horse can afford to have a good buggy."[16]

Naturally, the growth of the market stimulated carriage production. Whereas in 1850 the value of production was $18 million, by 1890 it had reached $102 million. This growth was only possible because the methods of production matured at an equally astounding rate. The census of 1850 lists 6,000 carriage building establishments; by 1870 this number had increased to 15,500. Similarly the amount of capital invested in the carriage industry increased during these

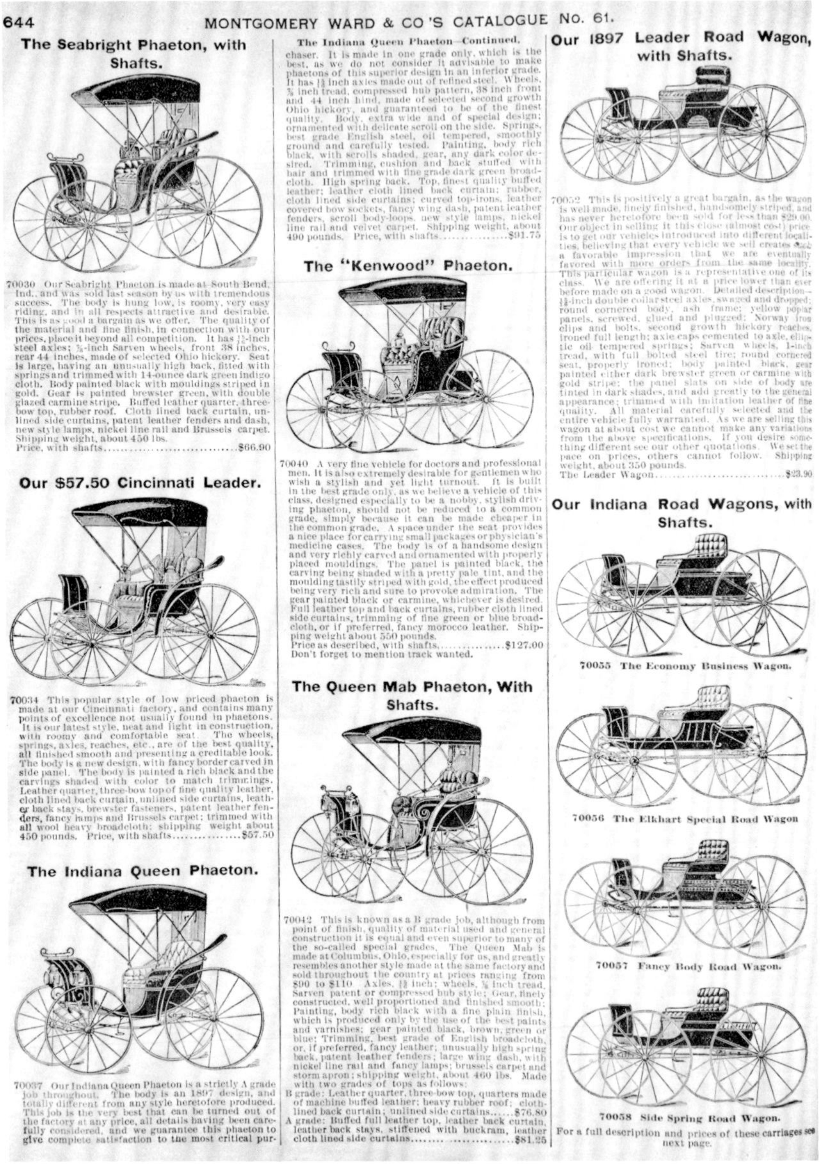

644 MONTGOMERY WARD & CO'S CATALOGUE No. 61.

The Seabright Phaeton, with Shafts.

70030 Our Seabright Phaeton is made at South Bend, Ind., and was sold last season by us with tremendous success. The body is hung low, is roomy, very easy riding, and in all respects attractive and desirable. This is as good a bargain as we offer. The quality of the material and fine finish, in connection with our prices, place it beyond all competition. It has 1⅛-inch steel axles; ⅞-inch Sarven wheels, front 38 inches, rear 44 inches, made of selected Ohio hickory. Seat is large, having an unusually high back, fitted with springs and trimmed with 14-ounce dark green indigo cloth. Body painted black with mouldings striped in gold. Gear is painted brewster green, with double glazed carmine stripe. Buffed leather quarter, three-bow top, rubber roof. Cloth lined back curtain, unlined side curtains, patent leather fenders and dash, new style lamps, nickel line rail and Brussels carpet. Shipping weight, about 450 lbs.
Price, with shafts..............................$66.90

Our $57.50 Cincinnati Leader.

70034 This popular style of low priced phaeton is made at our Cincinnati factory, and contains many points of excellence not usually found in phaetons. It is our latest style, neat and light in construction, with roomy and comfortable seat. The wheels, springs, axles, reaches, etc., are of the best quality, all finished smooth and presenting a creditable look. The body is a new design, with fancy border carved in side panel. The body is painted a rich black and the carvings shaded with color to match trimmings. Leather quarter, three-bow top of fine quality leather, cloth lined back curtain, unlined side curtains, leather back stays, brewster fasteners, patent leather fenders, fancy lamps and Brussels carpet; trimmed with all wool heavy broadcloth; shipping weight about 450 pounds. Price, with shafts.................$57.50

The Indiana Queen Phaeton.

70037 Our Indiana Queen Phaeton is a strictly A grade job throughout. The body is an 1897 design, and totally different from any style heretofore produced. This job is the very best that can be turned out of the factory at any price, all details having been carefully considered, and we guarantee this phaeton to give complete satisfaction to the most critical pur-

The Indiana Queen Phaeton—Continued.

chaser. It is made in one grade only, which is the best, as we do not consider it advisable to make phaetons of this superior design in an inferior grade. It has 1⅛ inch axles made out of refined steel. Wheels, ⅞ inch tread, compressed hub pattern, 38 inch front and 44 inch hind, made of selected second growth Ohio hickory, and guaranteed to be of the finest quality. Body, extra wide and of special design; ornamented with delicate scroll on the side. Springs, best grade English steel, oil tempered, smoothly ground and carefully tested. Painting, body rich black, with scrolls shaded, gear, any dark color desired. Trimming, cushion and back stuffed with hair and trimmed with fine grade dark green broadcloth. High spring back. Top, finest quality buffed leather; leather cloth lined back curtain; rubber, cloth lined side curtains; curved top-irons, leather covered bow sockets, fancy wing dash, patent leather fenders, scroll body-loops, new style lamps, nickel line rail and velvet carpet. Shipping weight, about 490 pounds. Price, with shafts.................$91.75

The "Kenwood" Phaeton.

70040 A very fine vehicle for doctors and professional men. It is also extremely desirable for gentlemen who wish a stylish and yet light turnout. It is built in the best grade only, as we believe a vehicle of this class, designed especially to be a nobby, stylish driving phaeton, should not be reduced to a common grade, simply because it can be made cheaper in the common grade. A space under the seat provides a nice place for carrying small packages or physician's medicine cases. The body is of a handsome design and very richly carved and ornamented with properly placed mouldings. The panel is painted black, the carving being shaded with a pretty pale tint, and the moulding tastily striped with gold, the effect produced being very rich and sure to provoke admiration. The gear painted black or carmine, whichever is desired. Full leather top and back curtains, rubber cloth lined side curtains, trimming of fine green or blue broadcloth, or if preferred, fancy morocco leather. Shipping weight about 550 pounds.
Price as described, with shafts.................$127.00
Don't forget to mention track wanted.

The Queen Mab Phaeton, With Shafts.

70042 This is known as a B grade job, although from point of finish, quality of material used and general construction it is equal and even superior to many of the so-called special grades. The Queen Mab is made at Columbus, Ohio, especially for us, and greatly resembles another style made at the same factory and sold throughout the country at prices ranging from $90 to $110. Axles, 1⅛ inch; wheels, ⅞ inch tread, Sarven patent or compressed hub style; Gear, finely constructed, well proportioned and finished smooth; Painting, body rich black with a fine plain finish, which is produced only by the use of the best paints and varnishes; gear painted black, brown, green or blue; Trimming, best grade of English broadcloth, or, if preferred, fancy leather; unusually high spring back, patent leather fenders; large wing dash, with nickel line rail and fancy lamps; brussels carpet and storm apron; shipping weight, about 460 lbs. Made with two grades of tops as follows:
B grade: Leather quarter, three-bow top, quarters made of machine buffed leather; heavy rubber roof; cloth-lined back curtain; unlined side curtains......$76.80
A grade: Buffed full leather top, leather back curtain, leather back stays, stiffened with buckram, leather cloth lined side curtains.......................$81.25

Our 1897 Leader Road Wagon, with Shafts.

70052 This is positively a great bargain, as the wagon is well made, finely finished, handsomely striped, and has never heretofore been sold for less than $29.00. Our object in selling it this close (almost cost) price is to get our vehicles introduced into different localities, believing that every vehicle we sell creates such a favorable impression that we are eventually favored with more orders from the same locality. This particular wagon is a representative one of its class. We are offering it at a price lower than ever before made on a good wagon. Detailed description—1⅛-inch double collar steel axles, swaged and dropped; round cornered body, ash frame; yellow poplar panels, screwed, glued and plugged; Norway iron clips and bolts, second growth hickory reaches, ironed full length; axle caps cemented to axle, elliptic oil tempered springs; Sarven wheels, 1-inch tread, with full bolted steel tire; round cornered seat, properly ironed; body painted black, gear painted either dark brewster green or carmine with gold stripe; the panel slats on side of body are tinted in dark shades, and add greatly to the general appearance; trimmed with imitation leather of fine quality. All material carefully selected and the entire vehicle fully warranted. As we are selling this wagon at about cost we cannot make any variations from the above specifications. If you desire something different see our other quotations. We set the pace on prices, others cannot follow. Shipping weight, about 350 pounds.
The Leader Wagon..............................$23.90

Our Indiana Road Wagons, with Shafts.

70055 The Economy Business Wagon.

70056 The Elkhart Special Road Wagon

70057 Fancy Body Road Wagon.

70058 Side Spring Road Wagon.

For a full description and prices of these carriages see next page.

From Catalog No. 61
Montgomery Ward and Company
Spring and Summer 1897
Chicago, Illinois

20 years: In 1850 a total of $8 million was invested in the American carriage industry; by 1870 the figure had increased to nearly $40 million.[17]

The growth of production also inspired changes in methods of marketing carriages during the nineteenth century. During the earliest years of the century most carriages in the United States were of the finest class of vehicle, usually purchased from importers who had acquired them from Europe. As the American industry developed and as the carriage was "democratized," methods of marketing diversified; by the end of the century carriages could be purchased from a showroom or repository, from a catalog, from a dealer, or directly from the manufacturer.

The repository developed rather early in the nineteenth century, when New York City-based manufacturers were faced with competition from coachmakers in both the countryside and neighboring commercial centers who would come to the city to show their wares. In an effort to remove this source of competition, carriage makers appealed to the Common Council to prohibit the "out-of-towners" from showing their products on the New York City streets. In order to sell their carriages, the visitors were forced to establish showrooms or repositories.[18] The showroom proved to be so successful that by the end of the century it was not unusual for major carriage manufacturers to have branch offices in several cities.

An individual did not, however, have to go to a showroom or the factory to purchase a carriage. He did not even have to leave his home, for by about mid-century, but particularly after the Civil War, the carriage catalog became a favored method of sales for both consumer and manufacturers, as attested to by the extensive number of catalogs available. Varying in size, style and glamour, they all advertised their products along with any options that were available to the consumer. All the pertinent information such as prices, warranties and mail-order procedures were also included. While some factories sold their wares only in this manner, others offered the mail-order alternative in addition to their factory showrooms. Some mail-order houses, such as Montgomery Ward and Sears, Roebuck and Company, included a selection of carriages along with the other products they sold.

Horse-drawn vehicles could also be purchased directly from the factory. There they could be chosen from stock items or made to a customer's particular specifications. Although an individual was sometimes allowed limited choices when buying from a catalog, such as paint color and wheel size, the possibilities were without limit when he sat down with a manufacturer to design a horse-drawn vehicle or to modify an existing style to suit his taste and purposes.

The development of alternatives to purchasing a vehicle directly from the factory obviated the need for the customer and carriage maker to be in close proximity, and the industry was no longer restricted to areas of high population density such as the East Coast commercial centers. Consequently the nineteenth century witnessed the slow yet steady movement of the center of the industry from the traditional urban centers of the eastern seaboard to the newer towns of the Midwest. In 1820 the center of the carriage industry was on the East Coast with 99% of production located there. This decreased to 88% of the business in 1850 and 57% by 1860; by 1880 only one-third of the carriage industry was located on the East Coast. This decrease corresponded to a steady increase in the fraction of production that was located in the northern Midwest. While in 1860 only 19% of the industry was in the northern Midwest, by 1880 59% of the horse-drawn vehicles produced in the United States came from this section of the country.[19]

There were several reasons for the movement of the industry away from the eastern seaboard to the northern Midwest. First of all, land was cheaper, so the factory, which required a more spacious locale than the small carriage shop, was better suited to the Midwest. Second, many of the raw materials used in production, particularly lumber and iron, were more abundant and

Vehicle Packed for shipping, "As it looks When Opened."
May 1907
The Vehicle Dealer
Publisher, Ware Brothers Company, Philadelphia, Pennsylvania

consequently cheaper in the Midwest. Finally, the general westward movement of settlers stimulated a tremendous rate of growth in this section of the country, which created a local market for horse-drawn vehicles. Thus, during the second half of the nineteenth century, as the center of the United States was moving westward, so also were the specialty and vertically integrated carriage factories.

By the end of the nineteenth century the dynamic growth of the American carriage trade had produced a diverse industry. The various methods of production employed in the manufacture of the domestic horse-drawn vehicle resulted in the availability of a wide variety of wagons, coaches and carriages. While factories generally mass-produced relatively inexpensive vehicles, the expensive custom-made carriage could still be obtained from exclusive carriage firms which employed designers. Further-more, the customer could purchase the vehicle of his choice in the manner most convenient for him; carriages were sold in branch office showrooms, through catalogs, or at factories located almost anywhere–small towns or big city centers–since by the end of the century the industry had virtually criss-crossed the nation.

While the themes of growth and diversity characterize the overall effect of industrialization on the process of carriage manufacture during the nineteenth century, the fine texture of these changes can best be appreciated by a closer look at the fabric of the industry. For this a survey of three particular carriage manufacturers is most useful. The three are Brewster & Company, a maker of fine, private vehicles; Studebaker Brothers Manufacturing Company, a manufacturer of mass-produced vehicles; and Abbot-Downing Company, best known for the production of solid, commercial vehicles. (Abbot-Downing Company was also known as Abbot, Downing & Company for a brief time in its history; for the sake of consistency, the former usage is used throughout this book.) Brewster, Studebaker and Abbot-Downing each began carriage building in the typical early craft tradition. During the latter half of the nineteenth century all three companies grew to be major forces in the industry, mavericks in production and trend-setters in design. Despite these similarities, however, the nature of the development of each of these companies, the details of its particular manufacturing methods and the character of the vehicles it produced were unique to that particular company. Brewster & Company catered to the needs of the upper classes and produced expensive vehicles of excellent quality and artistic style. In contrast, Studebaker wholeheart-edly embraced the ethos of industrialization and produced low-cost and dependable wagons and carriages. Abbot-Downing was best known for the sturdy and durable construction of its coaches, which were depended upon to open the West and conquer new frontiers, and to provide public transportation in rural areas in the eastern United States.

BREWSTER & COMPANY

Brewster & Company began as a small-town carriage shop born in the craft tradition. In 1804 James Brewster served as an apprentice to Charles Chapman of Northampton, Massachusetts. In 1810, he opened a carriage shop of his own in New Haven, Connecticut; there he ran a successful business for many years and created a strong local reputation, especially for dividing the work into separate departments. He taught carriage making to his sons James B. and Henry, who eventually opened shops in New York City. Henry opened a carriage shop in 1856 on Broome Street under the name of Brewster & Company; at about this time James B. took over the firm established by his father in New York City and sold horse-drawn vehicles under the name of J. B. Brewster & Company, in competition with his brother.

While Henry's business prospered and his vehicles gained an international reputation among the world's gentry, eventually the business of James B. found less success. When his business failed in 1895, the 80-year-old entrepreneur was approached by Cairn Cross Downey, George M. White and Henry M. Duncan, who proposed a "reorganization" of the company. A judicial hearing in 1908 exposed the fact that Downey, White and Duncan had persuaded an ailing James B. Brewster to sign over the use of his

Brewster's Carriage Factory, 25th Street, N. Y.

name. The new company attempted to pass off lesser-quality carriages under the prestigious Brewster label, displaying a portrait of James Brewster as well as the family coat of arms to convince customers that they were indeed producers of the first-rate Brewster carriage, while advertising their low prices with the statement, "Do not let the name Brewster frighten you away from us, even if you want only a low priced vehicle." The courts finally put an end to this hoax in 1908.[20] Luckily, Henry's business, Brewster & Company, did not suffer any irreparable damage from the antics of Downey, White and Duncan. In fact, it continued to thrive into the twentieth century under the leadership of Henry's son, William Brewster.

Brewster & Company of Broome Street produced vehicles that were well known for their excellence of design, quality of materials, and light though durable composition. They used hickory spokes, elm hubs, and whitewood panels, and were among the first to use steel to reinforce the wood, avoiding the need to rely on heavy wooden structural members for support, and thus creating the lightweight yet durable Brewster carriage that enjoyed a widespread reputation. Brewster & Company was awarded at the 1878 International Exposition in Paris both a gold medal and Henry Brewster was made a Chevalier of the Legion of Honor. Thereafter this company enjoyed one of the best reputations for carriage building in the world, attracting as customers the most prominent members of American society, including members of such notable families as the Vanderbilts, the Astors and the Rockefellers.

The appeal of vehicles made by Brewster & Company to the uppermost strata of society was maintained throughout its years in New York. Its dealings with the fashionable "Four Hundred" required "social understanding, tact and the ability to give service smoothly and expeditiously." A commitment to satisfying every need of the customer motivated the decision to maintain a special

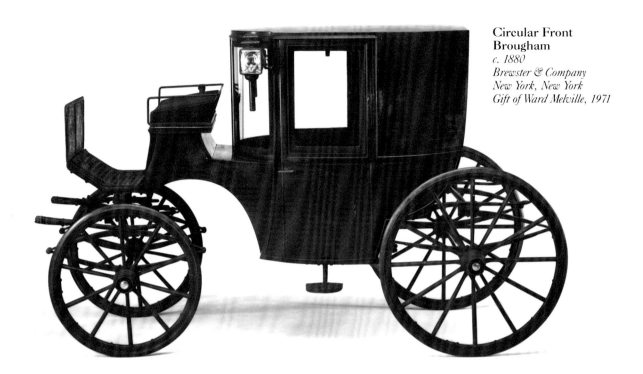

depot in Newport during "the season."[21] Accessibility also motivated the decision to keep the factory located in the rather expensive commercial hub of New York City. Although many orders could be made through the mail, quite often a Brewster customer would come to this New York City establishment and choose the fittings, for example, hoods, fenders, wheels, lights, and other accessories from a selection displayed in a catalog and create the particular vehicle that was desirable. A draftsman would develop a rather detailed illustration of the vehicle, including such particulars as crests and monograms applied to the panels, and this sketch would then be enlarged and painted in the color the customer chose, and serve as a schematic of the finished product.[22]

Although this method of sales appears to be quite traditional, Brewster & Company did actually employ modern industrial methods of manufacture; rather than totally embracing nineteenth-century industrial advancements or emphatically rejecting them in favor of custom production, Brewster & Company found a way of combining the two. By limiting the varieties of the parts of its vehicles, Brewster & Company was able to standardize much of the production process and assemble at a high rate, restricting most of the labor-intensive work to the final stages of production and finish. After the vehicle was assembled it would often be personalized according to the customer's specifications. Both exterior adornments and interior accessories decorated a Brewster & Company carriage. Some of the external ornamentation included monograms, and specified color combinations. Interior accessories included card cases, watches, extra stuffed cushions to satisfy such needs as the "lady [that] is very light and must sink into trimming;" mirrors; brass; carriage lace; silver or ivory fittings; umbrella holders; and glass racks. Brewster & Company would also often supply the other necessary accoutrements that a driver of such a vehicle would need to have on hand "just in case," including the formula for mixing the oil for lubricating axles, an axe, a wrench to fit precisely the axle nuts of the vehicle and extra washers, shackles and extra bolts.[23]

To produce the finest quality carriages,

Brewster maintained an excellent relationship with its labor force, particularly the skilled craftsmen. In this area Brewster & Company was every bit the innovator it had been in production and design. As early as 1870 the company began a system of profit-sharing, initiated by Henry Brewster's partner John Britton. On January 1, 1870, the Brewster & Company Industrial Association was inaugurated; its constitution stated that at the end of each fiscal year the employees could divide amongst themselves 10% of the net profit. The money was to be shared in proportion to wages.

Other improvements included the establishment of a board of governors and a board of control. The board of governors had the responsibility of making the rules and regulations for the shop. It was comprised of the chairman of the board of control and representatives of the shop at large. The board of control was made up of the members of each of the seven departments of the factory: (1) heavy smiths, light smiths and finishers; (2) wheelwrights and carriage makers; (3) jobbers, cleaners-off, filers and platers; (4) heavy trimmers, light trimmers and stitchers; (5) body makers and light body makers; (6) body painters and (7) carriage part painters.

All employees had the opportunity of being members of the association as long as they were skilled mechanics, employed by Brewster for at least six months, and at least 21 years of age. Those prohibited from membership included clerks, salesmen, porters, apprentices, cartmen and "persons under instruction in any department" as discerned by Brewster & Company. Membership in the association permitted voting

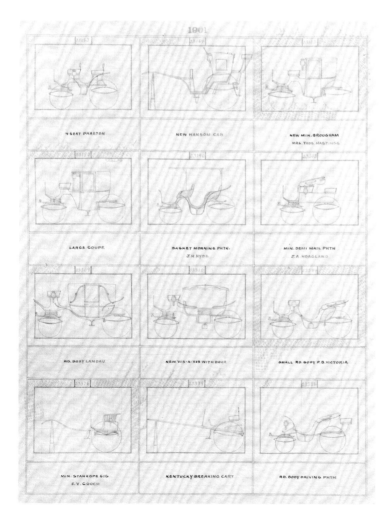

Album Leaf from
Designs for Carriages
1893-1904
Brewster & Company
New York, New York
Watercolor on paper
Photograph courtesy of
The Metropolitan
Museum of Art,
New York, New York
Gift of William Brewster,
1923

on the issues that came before it.

The success of the implementation of these innovations was noted in a trade journal that reported as late as 1891 that

The shop is run on a model plan. Employees are never discharged except on great provocation or for incompetency, and men who have grown old in the service and got too old to do work are pensioned and receive a stated sum per week. There is no tyranny, no oppression, and all hands honor and respect their head, and work together for the common good.[24]

Between the time that Brewster & Company first opened doors in New York and the time that it closed them in 1925, the company employed many innovations in production that had notable effects on the nature of the vehicle produced. By looking at the record of sales and comparing the nature of the product manufactured in 1858 to that made in 1886, these changes become evident. The business moved to New York City in 1856; by 1858 it was still rather young, and continued to employ many traditional techniques of production. In contrast, by 1886 Brewster had mechanized production. Furthermore, by this time the company was reaping the benefits it gained after securing its reputation at the International Exposition of 1878. In 1858, 251 vehicles were sold in New York; these vehicles were of 19 types, of which the coach and the road wagon were the most popular–69, or 28%, were coaches; 63, or 25%, were wagons. In the 30 years that followed, Brewster's production more than tripled with 848 sales for the year September 1886 through August 1887. Road wagons were still an extremely popular vehicle, comprising 24% of sales, followed by the phaeton, which had 16%, and the brougham, with 14%. The coach, which just 30 years earlier had been the highest-selling vehicle, accounted for only 1.3% of sales in 1886-1887. This figure reflects the increasing popularity of lighter vehicles; the coach averaged a weight of 1758 pounds, as compared with the 252-pound wagon or the 549-pound phaeton that were popular in the later period.[25]

The degree and nature of customization changed during this period. In 1858 almost every request for special items involved the addition of a surcharge on the price of the vehicle. This additional charge varied from 75¢ for altering curtains, to $65 for a new shifting top, and $75 for a harness; the average surcharge per customer was about $20. This was quite common at mid-century. In contrast, in 1886 the appearance of a surcharge in the ledger books was comparatively rare. Perhaps this was indicative of changing tastes and styles, but most clearly it was representative of changing

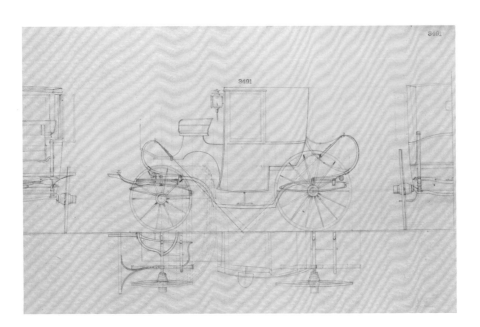

Album Leaf from Draftbook d'Orsay Drawings
c. 1890
Brewster & Company
New York, New York
Ink and watercolor on paper
Photograph Courtesy of The Metropolitan Museum of Art, New York, New York
Gift of William Brewster, 1923

Carriage: Brougham
c. 1890
Brewster & Company
New York, New York
Ink and watercolor
on paper
Photograph Courtesy of
The Metropolitan
Museum of Art,
New York, New York
Gift of William Brewster,
1923

company policy. In the earlier period such apparent essentials as poles, hubcaps, stable covers, lamps, slip linings, shafts, curtains and straps were not included in the base price of the carriage. Additions of the latter period tended to be more luxurious. For example, watches, extra stuffing, gold and silver hardware, glass racks, vanities and umbrellas were requested. Items specifically called for at mid-century were now included in the base price. In addition, as previously mentioned, by the fourth quarter of the century the Brewster vehicle was a symbol of prestige and wealth; the ornamentation on the carriage was therefore not to be spared, notwithstanding additional costs.

Clearly the vehicle of the later period sold at a much higher price than those of the earlier period; in 1858 the average price of a vehicle sold by Brewster & Company was $445.18. Thirty years later the average cost was $900.60, double the previous amount.[26] This increase, although coinciding with a trebling of production, at least in part reflected the increase in labor-intensity and cost of materials for custom work that Brewster's clients demanded.

Nonetheless, an increase of this magnitude in price during a time of industrialization is somewhat unexpected, for the mechanization of the work process traditionally lessens the cost of that which is produced, making it more affordable. The question therefore arises: Why did Brewster & Com-

pany not drop prices when it modernized production? The answer lies largely in Brewster's continued commitment to production of frequently highly customized quality vehicles. Though these labor-intensive work processes were not cost effective, the upper strata of society, which demanded custom work, were certainly able to afford the price increase. Industry thus influenced and improved elegant vehicles, such as those made by Brewster & Company.

In contrast to this peculiar method of production, Studebaker Brothers Manufacturing Company produced vehicles for the masses by implementing industrial techniques on a large scale, by employing machinery to facilitate the mass production made possible by the development of standardized and interchangeable parts, and by concentrating on products which could be mass-produced and sold at a reduced rate.

STUDEBAKER BROTHERS MANUFACTURING COMPANY

Studebaker was founded by Henry and Clement Studebaker in South Bend, Indiana, in 1852 with $68, as a traditional wagon shop. Employing the craft system of production taught them by their father, they produced three wagons in their first year. Ten years later their volume increased to 417, and by 1868 the company had incorporated

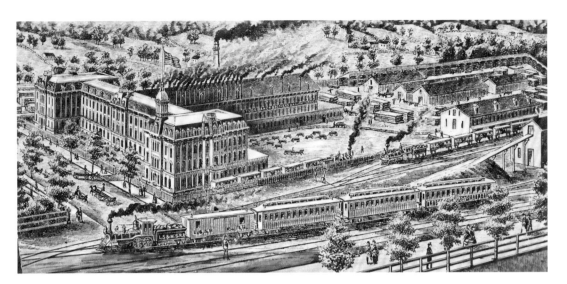

as the Studebaker Brothers Manufacturing Company, with $75,000 in capital stock. By 1872 the company produced 6,950 vehicles, grossed $691,000 in sales and maintained a labor force of 325, and its expansion continued rather steadily, with only occasional setbacks.

The initial success of the company was at least in part due to contracts from the United States government, which would remain one of the firm's most important clients. This business relationship began with orders for wagons needed for the Civil War. However, even after the war, and throughout the rest of the century, Studebaker was a favored federal supplier. The first government contracts challenged the young company, which had not been prepared for high-volume production; initially Studebaker was unable to meet the demand for military wagons and ambulances, but it expanded sufficiently during the war to be able to fill all orders. The effect of these contracts was far-reaching, for in addition to their profitability, which stimulated expansion, the war spread the Studebaker name across the nation.[27] The prestige of selling carriages to several United States presidents further strengthened Studebaker's already positive reputation. The Studebaker brothers, consistently strong Republicans, thus benefitted from their position as the favored federal suppliers of both the military and the administration.

Modernization and expansion of the company's facilities were necessary for its growth. The initial impetus for physical growth was the rebuilding of the factory, which burned to the ground first in 1872 and then again in 1874. (Fires were common to carriage factories. Listings of companies destroyed by fire were regular features in trade journals. The combustible wood fire forges and inflammatory varnishes and paints under one roof created an ideal situation for fires. Spontaneous combustion in the paint shop, due to rags and chemicals, was not unusual.) Growth continued; by 1893 the Studebaker factory in South Bend occupied 95 acres.

Within the Studebaker factory labor was divided along the lines of individual work processes. This method of factory organization enabled highly efficient production. Departments included a harness department, forging department, wagon department, blacksmith shop, engine room, electric dynamo and boiler room, wood shop, polishing and grinding room, painting and trimming room, seat and veneer department, machine shop, tin shop and more.

In addition to expanded facilities and modern methods of factory organization, a large part of the magic of Studebaker's success was its use in any given decade of the latest machinery–a continuing commitment to mechanization and innovation. In 1893 Studebaker boasted the use of 808 machines powered by 16 steam engines, which produced 1,550 horsepower per day.[28]

These machines facilitated the production of standardized parts, which made possible mass production, "insur[ing] a perfect fit of every part, and the consequent perfection of the whole."[29]

One example of Studebaker's applying the latest innovative processes to production was electric resistance welding. This procedure automated one of the most difficult tasks of blacksmithing, eventually replacing more traditional welding techniques. Along with applying technological advancements made in other industries and other carriage companies, Studebaker initiated several innovations of its own. The patents that were granted to this company include numerous wheel jacks, several machines for welding tires and boring hubs, a tire heating furnace and a tire cooler.[30]

During the second half of the nineteenth century Studebaker embraced a twofold philosophy of business that brought the company much success. Besides implementing the techniques of mass production, Studebaker remained committed to a philosophy of underselling the competition by reducing costs. With production per vehicle costs kept to a minimum to maintain high volume, Studebaker could afford to sell its buggy for $77.50 and its famous wagon for $60 in 1893. One of its most expensive vehicles was a three-spring phaeton, which sold for $250. The general popularity of the

Studebaker

Studebaker vehicle was due to its solid construction and affordable price and to the company's marketing system. Studebaker's commitment to low prices is evident in a discussion recorded in the company's *Minutes* of 1898, whereby the company was authorized to reduce the grade of work in order to reduce the cost of the World Buggy if the need arose.[31] In a further effort to keep prices down, Studebaker usually sold its vehicles "as produced," rather than made to order, although special parts of the vehicle, some accoutrements and even the horse could be purchased from the company.

Although the Studebaker brothers began a small business, building and repairing wagons and carriages and shoeing horses, in just a quarter of a century their company had become one of the largest manufacturers of horse-drawn vehicles in the nation. Similarly, the Abbot-Downing Company started as a small company and grew to command a national and international clientele.

Wagon Department Blacksmith Shop
1893
From Illustrated Souvenir of the Studebaker Brothers Manufacturing Company, *South Bend, Indiana*

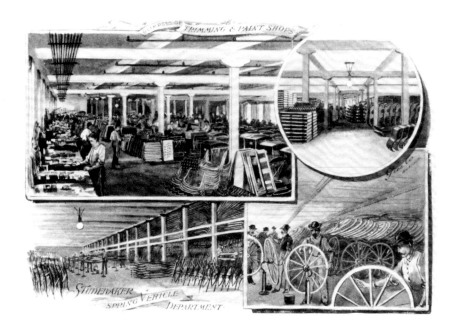

Glimpses of Trimming and Paint Shops and Spring Vehicle Department
1893
From Illustrated Souvenir of the Studebaker Brothers Manufacturing Company, *South Bend, Indiana*

Stanhope Buggy
c.1910
Studebaker Brothers Manufacturing Company
South Bend, Indiana
Gift of Joyce Barber, 1969

No. 6000 "Izzer" Stanhope
1913
From catalog No. 802, Studebaker Brothers Manufacturing Company South Bend, Indiana

Studebaker was best known for its well-built, moderately priced wagons and buggies. The "izzer" buggy became one of the country's most popular buggy models in the late nineteenth century. The name "izzer" was a clever adaptation of a rural colloquialism, "izzer," meaning something that is, in the present, as opposed to a "wuzzer," something that was, in the past.

ABBOT-DOWNING COMPANY

During the first decade of the nineteenth century Lewis Downing learned the craft of carriage making from his father and brother in their Lexington, Massachusetts carriage shop. In 1813 he moved to Concord, New Hampshire, and set up his own shop with $100 worth of tools and $60 in cash. During his first year of operation, Downing built pleasure wagons, and later, Concord buggies. He constructed all of the parts of the vehicle by himself, working only with hand tools, except for the iron work, which was fabricated by the inmates at a nearby prison. His vehicles contained no springs: the body was connected directly to the running gear.

In 1816 Downing expanded production by establishing a factory that employed between 10 and 12 men yet continued to rely upon traditional work processes. In that year Downing placed a help-wanted ad in the local newspapers that allows us a glimpse of the nature of his operation:

Lewis Downing, Has constantly on hand, opposite the Concord (Upper) Bank, small waggons which he will sell as cheap as can be bought, and warrant them to be well made of the best materials. A stout BOY, 16 or 17 years of age, wanted as an apprentice at the wheel-wright Business.[32]

This relatively moderate-sized operation was maintained for the next ten years, during which time Downing found himself increasingly turning to the production of heavyweight vehicles. Although he never ceased manufacturing private carriages and wagons, Downing's decision to concentrate on the production of heavier vehicles was largely due to the fact that they were more profitable.

By 1826 business was prospering enough to allow Downing to contract J. Stephen Abbot, a coach body builder from Salem, Massachusetts, to make three coach bodies. Apparently this association was successful, for soon after this Abbot became a partner. This partnership lasted for twenty-one years; in 1847 it was dissolved by mutual consent, and Abbot and Downing each opened independent shops, actually running them in competition with each other.

J. Stephen Abbot retained the original plant and Downing built a new one. Downing's two sons, Lewis and Alonzo, joined him; Abbot remained alone for five years until 1852 when he was joined by his son. In 1865 J. S. and E. A. Abbot and Lewis Downing and Sons once again merged; this merger coincided with Lewis Sr.'s retirement. Eight years later, in 1873, Abbot-Downing incorporated upon uniting with the formerly competing firm of Harvey, Morgan and Co. Numerous incorporations followed until 1932, when the plant doors were finally closed.

Throughout the nineteenth and into the twentieth century the productive process employed by Abbot-Downing evolved in response to the changing methods of manufacture available, and expanded to fulfill the public's growing demand for the company's products. Until about 1830 the process of production was centered about one craftsman's work. During the second third of the century, however, innovations of factory organization began to take hold. One such change was the division of carriage production according to particular work processes. In that year production was divided among the wood, smith and trimming shops. In addition to the craftsmen in these departments, other artisans with specialty skills were often called upon to perform particular work processes. One of these skills was turning and fitting axle molds.

When [L. Downing, Jr.] was old enough to drive a horse, he was continually being sent on errands for the company, and remembers especially when he used to be sent with axle molds made at the shop, over to East Concord, where they were turned and the pipe boxer fitted on by General Eastman and later for the same purpose to West Concord to be turned by Emery Burgess.[33]

The author of this passage goes on to state that this practice no longer continues. Rather, "everything of this kind is made and finished by machinery." Perhaps the employment of machinery permitted the vertical integration of the productive process, because by the 1870s the Abbot-Downing catalog claimed that "all parts–including axles and springs" were made within its works.

Nonetheless, within this modern

factory organization craftsmen trained in the traditional fashion continued to dominate production. At mid-century the labor force included "apprentices [who] were required to serve six years [in which] the father of the boy had to guarantee faithful performance of his duties until he reached his majority."[34]

The predominance of craftsmen did not preclude the introduction of machinery into the factory. Rather than resisting mechanization, craftsmen often welcomed the alleviation and simplification of tedious and difficult tasks. For example, the process of making wheels in 1813 was compared to production by machine at the close of the century thus:

[In 1813], by hand the seasoned plank was sawn into strips for spoke and felloe, by hand they were fashioned with the shave. By hand the tire was set. Now machinery does it all. A saw produces spoke and felloe-pieces as if by magic, as if by magic also a lathe turns out the finished part. A machine now prepares the hub and the box, and most recent of all, a machine now sets the tire.[35]

Furthermore, machinery did not detract from the individual attention given to each vehicle. In fact, Abbot-Downing advertised the custom work it performed. Although the concord coach was built in three standard sizes, it was stressed "every vehicle had distinguishing characteristics as decoration, fittings and construction varied according to the purchaser's special requirements."[36] A catalog circulated outside the United States promised to accept orders for any style of vehicle, including those not in the catalog "that the fancy of the purchaser may dictate."

The response to Abbot-Downing's method of integrating custom work with mechanization was clearly positive. In 1865 the value of production topped $360,000.[37] More than a thousand vehicles were ordered in that year, including passenger wagons, gigs, trade vehicles and coaches.

Throughout the history of Abbot-Downing the company continued to maintain many craft traditions while implementing new industrial methods of production. Although machinery assisted the work processes, craftsmen remained central to production. Production reached high volumes, but each vehicle was finished in accordance with the purchaser's particular specifications.

Abbot-Downing's concord coach was particularly well suited to a rugged environment. The exploration, settlement and subsequent exploitation of the West would not have been possible without such vehicles that could withstand the rough terrain of the frontier. The concord coach was built at the company's factory in Concord, New Hampshire, and shipped across the western frontier. One observer recalled the important role that this vehicle played in the days of the untamed wilderness:

It was the Concord coach which made staging possible over primitive roads and mountain trails. With only the European type of coach, staging in the West probably never would have succeeded.[38]

These vehicles were initially brought out West in the late 1840s by miners during the gold rush. In the years that followed, however, they were used for a variety of functions including expediting the transport of mail to the boomtowns as well as transporting people. Usually the coach was owned by a transport company and ran a regular route to deliver both people and mail.

The success of the concord coach was due in great part to its first-rate design and construction. Its superiority was applauded by one observer: "None else [but the concord coach] could be depended on, in the matter of springs and fittings, to pass us safely through all the variety of road we are likely to meet with."[39] The material was of the highest grade available. "The spokes, made from the finest ash, were literally hand picked and hand fitted to the rim and hub."[40] The wood for the body–stout oak–was seasoned to withstand any climate.

One of the most outstanding features

ABBOT, DOWNING & CO.

Concord Coach
c. 1875
Abbot-Downing
Company, Concord,
New Hampshire
Gift of Webster Knight II,
1962

of the concord coach was its suspension system of thoroughbraces. Thoroughbraces, or strips of steerhide, each about one-quarter inch thick, were piled to a thickness of about three inches and attached beneath the vehicle, from the front to the rear, and the body hung upon them. The swinging motion that resulted was a great improvement over the up-and-down motion of the conventional spring system, especially on the rough terrain of the West. The thoroughbraces absorbed road shocks. This method of suspension was not new, but its popularity had declined when wooden and metal springs were developed; Abbot-Downing renewed its popularity for overland vehicles. A great advantage of the leather thoroughbraces was that they could easily be replaced in the wilderness; spare leather braces could be kept on board the coach should repair be required; a broken wooden or metal spring was virtually irreparable and irreplaceable in the wilderness.

The impact of the concord coach was far-reaching, and was credited by some with opening up the West.

[The concord coach is] a heavy but apparently a light thing of beauty and dignity and life . . . eventually they found bumpy roads over the mountains and simultaneously with the pony express linked East and West. They blazed a trail that the railroads followed.[41]

It was upon these trails that Buffalo Bill rode to fame, at times in a concord coach– incidentally a favorite vehicle of his. Additionally, Mae Helene Bacon Boggs attested to the widespread use of this vehicle in California during her youth. She compiled a history of California for the years 1822–1888 and entitled it *My Playhouse was a Concord Coach*. There are references throughout this book to concord coaches in all their different roles, as mail coaches, miners' coaches and settlers' transport vehicles.

The demise of America's infatuation with the concord coach was attributed by some contemporaries to the rise of the railroad. Observers lamented that "when the trains came the coaches were left to rot in the rain."[42] Ironically, they continued, the railroad tycoons did not hesitate to follow the trails blazed by the frontiersmen in concord coaches. Nonetheless, concord coaches continued to be produced by Abbot-Downing and used in rural areas not reached by the railroad. Abbot-Downing also produced hotel coaches for use in rural and urban areas, and open wagons for sightseeing excursions. Nevertheless, when America rejected the coach in favor of the railroad

there were other frontiersmen ready to explore new wildernesses in a concord coach. Gold was discovered in Bathurst, Australia, in 1851, and the Australian gold rush followed. By 1855 various companies were in Australia operating the concord coach on a commercial basis. The popularity of the coach was great enough that Abbot-Downing found it advantageous to maintain two branch offices in Australia–one in Sydney and one in Melbourne–for at least 25 years. The coach was also popular in the rough wilderness of South America. An advertisement in the *Concord Evening Monitor* announced in 1899, "Stage Coaches for Peru: Abbot-Downing Have Girdled the Earth with 2,300 Coaches." Actually that number only includes those vehicles produced after midcentury. Estimates are that by the close of the nineteenth century Abbot-Downing had produced over 3,000 coaches.[43]

Although much of Abbot-Downing's national and international prominence was due to the popularity of the concord coach, the company in fact produced a wide variety of vehicles of both a private and public nature, for both personal and professional use. It manufactured these conveyances by implementing a process of production that blended traditional artisan skills with mechanized processes. Like both Studebaker and Brewster, Abbot-Downing was established in a time when the craft methods of production dominated and grew forward into the turbulent nineteenth century, when industrial innovations revolutionized traditional processes of manufacture.

The three companies discussed above illustrate the diverse nature in which the carriage industry grew and industrialized during the nineteenth century. Although they all made use of industrialized methods of production, the processes they employed, the products manufactured and the clientele

addressed defined three very different companies. Brewster focused on the production of the "custom" vehicle for society's elite and developed a system enabling the production of high-quality beautifully designed vehicles while incorporating many modern manufacturing techniques. Studebaker pursued a path of industrialization that emphasized the techniques of "mass production," which enabled it to become one of the largest carriage and wagon manufacturers in America. Abbot-Downing, like Brewster, employed industrial techniques primarily to assist the artisan, who remained central to the productive process; unlike Brewster, however, which was noted for the production of the light private vehicles of the gentry, Abbot-Downing applied the same principles of production to the manufacture of heavy, commercial vehicles that transported people both in the city and on the frontier.

Despite their differences, these companies did share at least one basic similarity—they all industrialized during the nineteenth century. The vehicles produced by each of them were manufactured by a vertically integrated carriage factory; all that was needed to produce the carriages, including supplies, tools and machinery as well as skilled and unskilled labor, was assembled in one location. Furthermore, the organization of each factory was departmentalized according to particular work processes. Finally, each of the companies did incorporate some degree of mechanization into the work process, although Studebaker may have exceeded Brewster and Abbot-Downing in the level of such innovation. As a consequence of the incorporation of industrial techniques, each company grew tremendously during the nineteenth century; whereas each began as a small locally-oriented craft shop, they all grew to command an international clientele.

Ironically, the final chapter in the history of each of these companies reflects the philosophy that motivated its development and stimulated its growth. During the first and second decades of the twentieth century, the automobile began to rival horse-drawn vehicles, finally asphyxiating the carriage-making industry. Brewster, which had chosen a path of modernization that maintained much of the traditional philosophy of the company–catering to an elite clientele and producing an extremely fine vehicle–turned down a tempting offer by Fisher to build bodies for General Motors because it would have had to abandon its custom methods for mass production, and sold out to Rolls Royce in 1925. Studebaker, which had embraced a very different and far more modern philosophy, choosing to undersell the competition at any cost even if this meant producing a vehicle of lesser quality than before, became an automobile manufacturer in the twentieth century. Abbot-Downing, which had dedicated itself to producing a solid vehicle, maintaining many craft traditions, added motorized trucks and fire equipment to its inventory in 1915; internally, production was limited to body construction and assembly and "each chassis was built to customers' specifications, as the old handicraft tradition was perpetuated into the new phase of Abbot-Downing's history."[44] Economic hardship, however, soon caused the abandonment of custom manufacture in favor of the standard truck model in 1919, and the plant doors were finally closed in 1932.

NOTES

1 Industrialization is a term adopted by historians to describe the changes in production dominated by the general acceptance of factory organization, mechanized mass production, machinofacture and standardized parts. The nature of this phenomenon has been the focus of much debate in the historical community. Some historians reject the nineteenth-century framework into which industrial change is often placed and the rather arbitrary distinctions between factory and shop. Certainly, characteristics of both the factory and the shop appear and continue to appear in one another; nonetheless, to facilitate discussion, distinctions have been drawn between the two modes of production. Similarly, because industrial processes were adopted before the nineteenth century, the chronological framework is also rather arbitrary. However, it was during that century that industrialization had its biggest impact upon the carriage industry, and has thus been adopted as the framework within which to focus this inquiry. For a discussion of historical development that gives a broader chronological framework to industrialization see Lewis Mumford, *The City in History*, (New York: Harcourt, Brace & World, 1961), and *Technics and Civilization*, (New York: Harcourt, Brace & World, 1934). Books providing useful background information include: Otto Mayr and Robert C. Post, eds., *Yankee Enterprise: The Rise of the American System of Manufactures* (Washington, D.C.: Smithsonian Institute Press, 1981); Nathan Rosenberg, *Technology and American Economic Growth* (New York: M.E. Sharpe, Inc., 1972); George Rogers Taylor, *The Transportation Revolution: 1815-1860* (New York: Harper and Row, 1951).

2 "Subdivision of Labor in Carriage Building," *The Hub*, 17.4 (1875): 122.

3 F. L. Pope, "Uses of the Electric Motor," *The Hub*, 30.4 (1888): 280.

4 *United States Census of Manufacture*, 1820.

5 *New Hampshire State Census of Industry*, 1870.

6 Ezra M. Stratton, *The World on Wheels* (New York: Ezra M. Stratton, 1878) 422.

7 Stratton 480.

8 "Representative Carriage Factories," *The Carriage Journal*, 1.1 (1963):24. [Reprinted from *The Hub*, 26.1 (1884)].

9 *G. and D. Cook and Co.'s Illustrated Catalogue of Carriages* (New Haven, Conn.: Baker and Goodwin, 1860) 5.

10 Stratton 467.

11 "Hope Ahead for the Skilled Carriage Painter," *The Hub*, 30.2 (1888): 113.

12 Stratton 467.

13 "Does Machinery Pay?" *The Hub*, 17.1 (1875): 5.

14 "The Deadwood Mail," *Concord Evening Monitor*, July 20, 1895.

15 G. A. Thrupp, *The History of Coaches* (London: Herlog and Endean, 1877) 134.

16 "Subdivision of Labor in Carriage Building," *The Hub*, 17.4 (1875): 122.

17 The data included in this paragraph were obtained from the *Twelfth Census of the United States Taken in the Year 1900. Manufactures, Part IV–Special Reports on Selected Industries.* (Washington: U.S. Government Printing Office, 1902). A note of caution should be sounded with regard to these statistics, as they tend to vary in accuracy according to the skill and honesty of the reporter, with a further margin of error in the process of transcription and recording. The numbers themselves should therefore be considered with an appropri-

ate amount of skepticism; their real value lies in the trends they depict, which substantiate the growth of the industry indicated by other sources.

18 Stratton 428.

19 These statistics come from the United States Census of Manufacture for the years 1810, 1820, 1850, 1860 and 1880. The caution described in footnote 17 above is also applicable to the statistics presented here.

20 Supreme Court, Brewster and Co., Plantiffs Against Cairn Cross Downey, Henry M. Duncan and George M. White, Defendants (New York: C. G. Burgoyne, 1908) 81.

21 Paul Downing, "The Brewsters and Their Carriages," *The Coachman's Horn*, 1.1 (1961): 20.

22 This process of manufacture was determined after studying pencil sketches and artists' paintings of Brewster vehicles at The Metropolitan Museum of Art, New York City. See also William Louis Gannon, "Carriage, Coach and Wagon: The Design and Decoration of American Horse-Drawn Vehicles," diss., University of Iowa, 1960.

23 Ledger sheet from Brewster & Company's account books, located at the New York Public Library, 9/21/1886 through 9/21/1887.

24 "James Brewster," *Varnish*, 4.9 (1891): 398.

25 These statistics were culled from the Brewster & Company account books of 1858 and for the year 9/21/1886 through 9/20/1887, located at The New-York Historical Society and the New York Public Library respectively. A certain amount of caution should be used when considering the actual data; like the statistics referred to in note 17, their importance lies more in the trends they depict than in the actual value they represent.

26 These prices were determined by averaging the prices of Brewster vehicles for 1858 as recorded in the Brewster & Company account books, currently in possession of The New-York Historical Society, and by a similar process applied to 1886-87 records at the New York Public Library. The change was described in "The Brewster and Co. Spring Opening, 1875," *The Hub*, 17.2 (1875): 48.

27 David A. Hounshell, *From the American System to Mass Production, 1800-1932: The Development of Manufacturing Technology in the United States* (Maryland: Johns Hopkins University Press, 1984) 147.

28 *Studebaker Souvenir Book, 1893*. Located in the Carriage Reference Library of The Museums at Stony Brook.

29 Hounshell 147.

30 Hounshell 147. See also the manuscripts in the Studebaker file in the Carriage Reference Library of The Museums at Stony Brook.

31 "Minutes of the Executive Committee, Studebaker Brothers Manufacturing Co.," June 13, 1898. Located at National Studebaker Museum, South Bend, Indiana.

32 Included among trade cards in the collection of Abbot-Downing Miscellany, New Hampshire Historical Society.

33 "Sixty Years," *Concord Evening Monitor* (May 3, 1897).

34 Edwin G. Burgum, "The Concord Coach" (1939), included amongst newspaper clippings in the collection of Abbot-Downing Miscellany, New Hampshire Historical Society.

35 *Concord Evening Monitor* (July 20, 1895).

36 Harry N. Scheiber, "Very Strong and Heavy: The Abbot-Downing Co., 1813-1928," in the Carriage Reference Library of The Museums at Stony Brook.

37 "Comparative Table Sales," in Abbot-Downing Manuscripts, New Hampshire Historical Society.

38 Burgum.

39 E. Daniel Potts and Annette Potts, *Young America and Australian Gold: America and the Gold Rush of the 1850's* (Queensland, University of Queensland Press, 1974) 105.

40 Burgum.

41 *Concord Daily Monitor*, n.d. Included among newspaper clippings in the collection of Abbot-Downing Miscellany, New Hampshire Historical Society.

42 *Concord Daily Monitor*, n.d.

43 *Concord Evening Monitor* (February 13, 1899).

44 Harry N. Scheiber, "Coach, Wagon and Motor Truck Manufacture, 1813-1928: The Abbot-Downing Company of Concord." New Hampshire Historical Society, 19.

A HARMONY OF PARTS:

The Aesthetics of Carriages in Nineteenth-Century America

Merri McIntyre Ferrell

Berlin Coach (*detail*)
c. 1780
Maker unknown, France
Purchase, 1953

A carriage is a complex production. From one point of view it is a piece of mechanism, from another, a work of art.[1]

Prior to the nineteenth century, carriages used for personal transportation by the wealthy and elite were expressions of applied ornamentation rather than form. The absence of technology and mechanical means of fabrication resulted in the manufacture of elaborate but cumbersome vehicles. From the mid-sixteenth century to the second quarter of the nineteenth century, horse-drawn vehicles were made in artisans' shops, and production was limited. The labor intensity required to construct vehicles made them expensive–excepting utilitarian carts and wagons designed for heavy work. Private conveyances were owned only by the very wealthy. Those members of the European aristocracy who could afford carriages took advantage of the opportunity to display their wealth and social position in mobile form. Royal patrons commissioned architects, artists and decorative artists to design and build carriages.[2] Ornament was profuse on state and ceremonial carriages. Structural elements such as spokes and pillars and, as exemplified by fanciful carousel sleighs, even entire vehicular bodies were sculptural. Although highly decorative, these vehicles were hung on relatively primitive and ineffective suspension systems, such as whip springs or leather braces hung from rigid pillars, and were generally uncomfortable for passengers. They were also extremely heavy and elongated, making them difficult for horses to draw even at a walk. Difficulty in draft was increased by the extreme difference in the diameter of low front wheels, which were designed to pass under the carriage in the act of turning, and the very high diameter of the rear wheels, which were intended to cover more ground when in motion.

Toward the end of the eighteenth century, improvements were made in technology, and carriage design. Greater consideration was given to technical improvements; the aim of vehicular design was directed toward improving the comfort of the passengers and reducing the elements that imposed greater exertion on the horse. "At this time, indeed, the old flamboyant ornamentation had all but disappeared from the carriages, which were in process of taking on the appearance they largely retain to this day."[3] The character of late-eighteenth-century vehicles was recorded in William Felton's *Treatise on Carriages*, first published in 1794. Although some ornamentation appears on the vehicles illustrated, the designs were more rational in terms of mechanical function. It can be seen that changes in cultural tastes were expressed in many types of artifacts, including carriages.

During the colonial period in America, horse-drawn vehicles were imported from England; those vehicles made domestically were English in character, based on designs borrowed and adapted from Felton's publication or similar design books. Like the decorative arts and architectural styles of the period, vehicular types were based on European precedents, modified by the lapse in time of communication between the continent and the colonies or the limitations of individual shops. Detailed descriptions of vehicles made during this time are virtually non-existent; trade journals or manufacturers' catalogs documenting the method of production or

Sleigh
c. 1780
Maker unknown, France
Gift of James Keillor, 1952

the appearance of vehicles were not published. Surviving examples of vehicles are extremely rare.

Domestic production of all sorts of goods was encouraged after the Revolutionary War, and the production of American-made carriages was stimulated. An example of a vehicle used during the early national period in America is a phaeton or four-wheel chaise in The Museums collection that belonged to General Peter Gansevoort of Albany, New York. Although the origin of manufacture is unknown, it was probably made in New York about 1780; it retains European characteristics, especially those recorded by Felton and other eighteenth-century carriage designers. The wheels are comparatively high and the body is hung on whip springs. The rear, ogee pillars give the body a graceful line. The body is articulated by half-round molding; the ends of the fore and hind panels terminate in carved scrolling in high relief. The gear is chamfered extensively; upright members are turned. The pump handles projecting from the rear of the carriage are scrolled and terminate in carved rosettes. Carving is accentuated with decorative paint. The elements of the gear are notably heavy, especially the perch.

Although the vehicular types of the late eighteenth and early nineteenth centuries continued to reflect European prototypes, specific American types emerged, such as the coachee, chariotee and pleasure wagon.

Gig (detail)
c. 1670
Maker unknown, Italy
Gift of Ward Melville, 1971

From The Nobleman
and Gentleman's
Director and
Assistant in the True
Choice of their Wheel
Carriages
1763
Publisher, A. Webley
London, England

Of these, the pleasure wagon is the least derivative and best reflects the new democratic spirit in vernacular design. These practical vehicles could be used for personal transportation or as working vehicles for hauling light freight. Fabrication by the hand and by the eye is evident in their construction. The bodies are curved and reinforced with wooden ribs. The application of the draw knife is visible in the chamfering of the gear members. Structural members were accentuated with paint applied in bold and contrasting colors. Ornamentation, when present, was carved onto the structural members, not superficially applied.

In the early nineteenth century carriage making continued in small shops. Many of the apprentices who trained in these shops became leaders in the carriage industry in the later part of the century. American carriages of the late eighteenth and early nineteenth centuries were transitional; specific national characteristics, such as lightness of construction, had not yet appeared. How they evolved to the level of "perfect expression of form" was summarized by George Houghton, a spokesman for the carriage industry, in a paper delivered to The New-York Historical Society:

Hereafter the vehicles gradually lose their character as representations of English aristocracy, and reflect with growing prominence the ideas and needs of American democracy. I am of course tempted to overstep the bounds of my subject, and show something of the later chapters of American Coach-building: how we long depended on the English makers for our most fashionable vehicles for gentlemen's use and for sporting; how France then contributed her quota in the way of vehicles specially adapted for ladies and park driving; how America copied, adapted and improved the imported types, and originated many new ones of her own; and how, in these latter days, New-York representatives of the art have thrice won the highest possible honors at World's Exhibitions. . . . and even those American types of wheeled Democracy; the buggy, the buckboard, the sulky and the Rockaway, will lack half their significance unless contrasted, hub to hub, with a full-liveried Colonial coach.[4]

The simplification of horse-drawn vehicular design was an immediate result of the industrial revolution. Industry had the potential to reduce the cost of items and to deliver intelligent designs to the masses at prices they could afford. Mechanical reproduction could guarantee standardization and

Four-wheel Chaise
c. 1780
Maker unknown,
United States or France
Gift of Johnstown
Historical Society,
Johnstown, New York, 1955

Four-Wheel Chaise
(detail)

precision. Experimentation with new materials was practiced. Ferro-ventrous structures such as the London Crystal Palace (1851) and the New York Crystal Palace (1853) were erected. Laminated and steam-bent wood were used for the mass production of furniture by Thonet and Belter, as well as in the construction of carriages.[5] Ornamental porches, as well as carriage dash and seat rails, were made of cast iron. Numerous international exhibitions celebrated the creations of the newly-mechanized world. The first of these exhibitions was organized by Prince Albert and held in 1851 in London's Crystal Palace, designed by Joseph Paxton. It displayed everything from garden furniture to horse-drawn vehicles. Contemporary commentary on the profusion of products and designs exhibited observed, "beauty of effort and decoration are no more a luxury in a civilized state of society than warmth or clothing are a luxury to any state: the mind, as the body, makes everything necessary that it is capable of permanently enjoying. Ornament is one of the mind's necessities which it gratifies by means of the eye."[6]

The preference for applied ornamentation in diverse styles was evident in many types of artifacts in the nineteenth century–from architectural facades to decorative arts. An exception to this trend was the horse-drawn vehicle.

By 1850 all types of American vehicles had begun to assume attributes that gave them their distinct national style. Abundant resources of timber, especially hardwoods such as hickory, and the application of machines in the various areas of production influenced the light construction of American carriages.[7]

American carriage manufacture was influenced by an emphasis on use–the prescribed function of a vehicle–and an adherence to the principle that vehicular design should respond to the requirements established by a diverse, independently-minded and increasingly mobile population. As described in an 1860 account, "In this country of 'magnificent distances,' we are all, more or less, according to the requirements of either business or pleasure, concerned in the use of riding vehicles."[8]

While in the midst of realizing the

Pleasure wagon
c. 1820
Maker unknown,
probably United States
Gift of Johnstown
Historical Society,
Johnstown, New York, 1955

From The Crystal
Palace Exhibition,
Illustrated Catalogue
London
1851
The Art-Journal
Special Issue,
(Reproduction by
Dover Publications, Inc.
New York, New York, 1970

THE INDUSTRY OF ALL NATIONS.

Another description of American carriages is copied from a single horse phaeton, manufactured by Mr. WATSON, of Philadelphia. One peculiarity we notice in it, is the unusual size of the fore- wheels compared with the hinder, so contrary to the practice of our carriage-builders, but there is no doubt this causes it to run easily. The body of the vehicle seems very light in its construction.

A wicker GARDEN-CHAIR, contributed by Mr. TOPF, of New York, possesses much novelty, and no little taste, in its ornamental design.

The TABLE-CLOTH called the "Ardoyne Exhi- bition Pattern," is another of the beautiful fabrics of Mr. M. ANDREWS, of Belfast. It was designed, in competition, by J. Mackenzie, of the Belfast School of Design, who richly merited the prize he obtained for a composition so excellent.

improvements made in the construction of horse-drawn vehicles, carriage makers continued to strive for ways to upgrade mechanically and aesthetically their products and system of manufacture, as noted by a contemporary observer:

It goes without saying that the growth of so vast an industry has called into action, or rather developed the place for, an immense array of skilled artisans of a very high order of ability, and has, also, to a remarkable degree, incited the inventive genius, and caused the production of an endless list of devices for the improvement of vehicles and all their various parts. The vehicle of today, whether a pleasure carriage or a heavy truck for business purposes in our great cities, is in striking contrast with those in use but a few decades ago.[9]

The design requisites established by mid- and late-nineteenth-century carriage makers were harmony, proportion and balance.[10] The mechanics of carriages dictated their form; ornamentation was subordinate to that form. "The most beautiful forms are always the simplest–the most obnoxious to good taste, those that are overladen with ornament."[11] The mobility and consequent visibility of carriages required that their form be pleasing; the application of this principle was carried out by American carriage makers.

European carriage makers were more traditional and regarded the mechanized carriage factories in America as indicative of a lack of skilled labor. They were especially critical of the lightness of American vehicles, exemplified by buggies and road wagons. A report in the English publication *The Carriage Builders' Art Journal* was quoted in 1860 in *The New York Coach-Maker's Magazine:*

Among the great variety of objects of manufacture, there are none more capable of displaying taste and judgment than carriages; and none are subjected to a more severe test of public opinion, being confined to no locality or class. And while we cheerfully give our Transatlantic brethren every credit for their skill and good judgment in every other article of manufacturing enterprise, we must be allowed to except that of Carriage Building, upon which public opinion has set its seal. That they are marvelously light, extremely well put together, most ingeniously constructed and designed for their purpose, we readily grant; also

that the draught is so light as to lessen fatigue to the horse, is a valued point gained. Yet these advantages are gained by an absence of quiet good taste, a want of proportion, a deficiency of that solid, substantial quality, that forms the characteristic of an English carriage. If we are to define the matter, then, we take these spider-looking vehicles, having enormous wheels, with scarcely any difference in the height, before or behind, with spokes less in diameter than an ordinary broom handle; against this the body is extremely shallow in height, giving it a low, squatty appearance, rather a vulgar aspect.[12]

The criticism of American carriages, especially as related to light construction, was reversed at the Centennial Exhibition held in Philadelphia in 1876. In the numerous exhibitions that followed, American vehicles not only won prizes, but received extensive commendations in European publications such as the English *Reports on Carriages in The Paris Exhibition:*

American carriage-builders . . . have made great progress during the last few years, not only in specialties of light vehicles, such as Wagons, Buggies, and Trotting Carts, for which they stand unrivalled, and in the designing of which they show great discrimination, by proportioning the various component parts, to obtain the maximum of strength with the minimum of weight in the material employed, but also in the heavier description of European carriages, of which a leading New York firm has a very fine exhibit, the Drag, Brougham, and Landau being especially worthy of notice.[13]

Harper's Weekly described the progress of American carriage manufacturers as displayed in the Centennial Exhibition: "Up to the year 1865 the superiority of English and French carriages was generally acknowledged, but since then the United States has taken a stride, and placed itself in the foremost rank."[14]

American carriage manufacturers were, by their own admission, indebted to continental prototypes; they had borrowed the best styles and methods of fabrication and, through the application of ingenuity and technology, had established themselves as creators of equal if not superior vehicles.

The adaptability of our people is no where more distinctly evident than in the building of our better types of carriages. We have united the simple and practical designs of the English builders with the perfection of detail that was developed by the French artisan, and our native forests and skilled labor have added the best material and workmanship; with the result that our carriages have been recognized as superior to those of any other country.[15]

The principle of American carriage design was to express purpose of use. The purpose of display evident in eighteenth-century aristocratic vehicles ran counter to the purpose of use and the reduction of components to functioning parts. A carriage was a composite object, combining a variety of materials such as wood, paint, leather, metal and textiles; it was also a summation of diverse skills. From the most formal coach to the most lightweight, inexpensive buggy, a horse-drawn vehicle had to accommodate passengers and withstand the strain of horse-power drawing it and the impact of the various road surfaces it traversed, and combine all these requisites in such a way as to convey balance, harmony and proportion–

regardless of the changing tastes imposed on or implied in vehicular styles or of the variety of styles.

The design determines the character of the vehicle. No amount of paint or metal work will alter the relation of the various parts to one another, nor change the lines. There are certain laws which should govern the design of all types of vehicles, and it depends upon the degree of strictness with which these laws have been observed in creating the composition whether the design is good, bad or indifferent. These laws are proportion, simplicity and harmony.[16]

In addition to general requisites for carriage design, specific qualities were assigned to each style or combination of styles. "Every vehicle should truthfully convey to the eye the purpose for which it is intended: i.e. a road wagon by its form of construction should indicate that lightness is desired; a brougham must be made so that solidity and comfort are suggested as being the characteristic."[17] Formal or informal, or whether designated for urban or rural use, or for pleasure,

Coupé Rockaway
1871
A. S. Flandrau
New York, New York
Gift of Franklin Joseph, 1962

The rockaway was a distinctly American type of vehicle that evolved from the coachee and the germantown, vernacular adaptations of the coach. Developed in Jamaica, Long Island, around 1830, the rockaway became a popular enclosed family vehicle, featuring a roof that extends over the driver, who could be the owner or a hired coachman. The extremely light structural members, especially the wheels, pump handles and axles, are typical of the modifications leading to reduction of weight and overall lightness characterizing American vehicles made during the second half of the nineteenth century.

sport or work, a vehicle should convey its function.

The relationship between the appearance of a vehicle and its use was fundamental:

No carriage can really look well if it does not look serviceable. If for heavy work, it must be strong and appear strong, without appearing heavy or unwieldy. If for light work, it can hardly appear too light; for it is so generally known that delicately made carriages may be very strong, that the light appearance does not convey an idea of weakness. If the carriage be intended to make a rich display, it will be most beautifully finished when most richly decorated. If to be used by persons of modest and retiring habit, it will be most beautifully finished when it exhibits great simplicity, or an entire absence of decoration. A royal state carriage, and a simple coupé are both beautiful, though widely different. The simple vehicle will usually be the most elegant. [18]

Confronted with and responsible for a rapidly growing industry and a proliferation of styles, members of the carriage industry recognized the need to publish information on the trade as a means to record activity in the business and to instruct carriage workers on ways to perfect their craft. A level of excellence had been achieved in their products; to ensure quality and establish standards, techniques had to be shared with other workers. Using the new media of popular literature, members of the carriage trade documented advancements in the field in various periodicals and manuals.

In 1858 Ezra M. Stratton, who had served as an apprentice to Charles Townsend of the carriage shop of Platt and Townsend in Connecticut from 1824 to 1829, and had opened his own establishment in New York City in 1836, began publishing a trade journal entitled *The New York Coach-Maker's Magazine*. This magazine featured articles on eminent carriage makers, techniques of various skills applied to carriage construction, illustrations of fashionable vehicles of every description, and news of the trade. Some articles were theoretical or pseudo-philosophical, bearing titles such as "Drawing, as Connected to Painting," "The Dignity of Labor" and "The Mechanic as an Intellectual Being." This and other trade publications demonstrated the degree of

self-confidence and self-consciousness possessed by carriage makers.

The New York Coach-Maker's Magazine continued to be published until 1871, when it was absorbed by *The Hub*, another trade journal devoted to the carriage industry. George W. W. Houghton became its editor in 1869 and, upon the merger of the two publications, became a spokesman for the carriage industry. Like *The New York Coach-Maker's Magazine*, *The Hub* offered informative articles on each subdivision of the carriage shop or factory. It continued to publish illustrations of currently popular styles of vehicles, accompanied by technical drawings, descriptive text and details of construction, finish and trim. It offered patterns for ornamental painting or monograms that could be traced and applied to business wagons and private vehicles. Diagrams for trimming, methods of constructing carriage bodies and descriptions of ironwork were also explained. A sample index of *The Hub* from the 1880s documents the division of skills required to make a carriage: "Wood Shop," "Smith-Shop," "Paint-Shop" and "Trimming-Shop." The other interests of the industry were addressed in "Office, Correspondence and Trade News," "Labor's Voice," and the more entertaining section known as the "Dinner Hour." From time to time, the journal offered cash prizes for articles submitted on topics such as "How can the Present Method of Carriage-Painting be Improved?"

As circulation increased and the magazine expanded, it carefully recorded news of the industry with monthly installations of trade gossip, advertisements, reports of national and international exhibitions including carriages, companies lost to fires and national statistics recording the activities of firms in all regions of the United States. It was dedicated to the advancement of the industry, with an eye not merely to the increase in profits on a national level, but also to promoting the skills of those currently associated with the business. Recording its own objectives, *The Hub* published a statement on the contribution of trade journals: "Not only do they lend dignity and importance to the trade in the eyes of the world, but they afford to the trade itself a means of ready inter-communication, the valuable results of which are daily visible."[19]

George Houghton was not directly associated with carriage building, but his interest and contributions were paramount. Among his accomplishments were his advocacy of and active membership in the Carriage Builders' National Association, established in 1872. He also promoted the formation of a technical school for carriage workers. The apprenticeship system had served to train young workers but was not sufficiently comprehensive to give them an understanding of the traditional methods they learned, nor did it provide them with a foundation on which to base new methods, other than by trial and error. The high standards established by carriage makers and those demanded by the public required continuity.

Technical schools in Europe had proved to be successful. The purpose of these schools was "to train the ambitious artisan, as that he may attain the highest excellence in his calling, whether it be carriage-making or any other trade involving manual dexterity combined with esthetic forms . . . for their tendency is to elevate the standard of excellence in any industry, and thus add to the dignity of both employer and employed."[20] By promoting aesthetics in industrial arts, technical schools enhanced the national product.

Technical schools for the teaching of artisans are essential to the progress of American industrial art. The time has certainly arrived when America should cease to be dependent upon foreign production of beautiful works in any and every department of industry; when American youths should have the facilities for learning how to produce beauty which German, French, and especially English youths have in technical schools. It may be affirmed that in every kind of mechanical work, however inferior its character among the arts, the mechanic who can design as well as do his work, is worth more to himself and to his employer than one who can only follow a director or do that which he has seen done. If American industrial art is to rank with that of European countries, it can only be by having educated artisans.[21]

The establishment of the Technical School for Carriage Drafting and Construction was proposed at the annual meeting of the Carriage Builders' National Association in 1879. An executive committee was formed in 1880, comprised of John Britton, master draftsman of Brewster & Company, New York; Wilder Pray; William Fitz-Gerald, carriage trimmer of Newark, New Jersey; and George W. W. Houghton. Financial support was contributed by members of the industry, many of whom were members of the Carriage Builders' National Association. The first classes were held at The Metropolitan Museum of Art, located in New York City on First Avenue and East 68th Street (both the school and the museum moved to East 34th Street in 1881). John D. Gribbon, formerly employed by Brewster & Company, was appointed the chief instructor. The prospectus of December 21, 1880, listed the following areas of study:

"Linear designing, including scale and full-size drawing
Geometry applied to Carriage Construction, including the principles of the 'French Rule'
Carriage Body-making
Construction of Carriage Gearing Wheel-making and Principles involved in the Suspension of Carriages."[22]

The majority of students who attended the school were employed in the carriage trade. Their attendance demonstrated their desire to improve specialized crafts through education. In addition to the classes offered, students were encouraged to attend lectures delivered by members of the industry on pertinent topics, such as the series given by Mr. F. B. Patterson on "The Ancestry of the Modern Pleasure Carriage," which traced the evolution of vehicular styles and categorized them in discrete groups.[23] The executive committee of the Carriage Builders' National Association devoted considerable time to enhancing the school's curriculum. In addition to sponsoring lectures, they established a fund for a special library that included rare and contemporary books on carriage history and technology and prize drawings of carriages.

In 1884 the school expanded through a "Chautauqua" system of correspondence classes so that it could accommodate interested members of the trade who were unable to attend the classes in person. Select

Metropolitan Museum of Art Technical Schools,

Corner First-Avenue and 68th-street, New-York.

March 15, 1881.

DEAR SIR:

We have the pleasure to announce that on the evening of Friday, March 18th, the second of a series of Lectures on subjects pertaining to Carriage Building will be given before the Class in Carriage Drafting and Construction, at the Technical School, corner First-Avenue and 68th-street, New-York, by Mr. F. B. Patterson, of New-York, on "The Ancestry of the Modern Pleasure Carriage," a subject to which Mr. Patterson has given special study. It will be profusely illustrated by diagrams.

The following "Genealogical Tree," illustrating the development of the principal families of Modern Pleasure Carriages, will be found useful to those who attend, in following the lecturer:

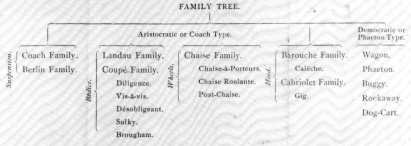

The School Building is easily accessible by means of the East-Side Elevated Railways, and its location and that of the nearest railway stations are indicated in the accompanying

lesson plans were published in *The Hub*. In 1885, the school moved from its location in The Metropolitan Museum, "there not being sufficient room in the premises for their accommodation,"[24] and relocated at the Young Men's Institute at 222 Bowery.

Through donations of books and materials and financial support, the carriage industry invested in its future by subsidizing the school. The significant role of the industry in the advancement of education differed from the support given it in other countries.

"Thus in France, and England, and America, the same means is employed to further progress and development of the carriage trade, and to arouse the intelligence and energy of those engaged therein, but with this characteristic difference: —in France, it is the government [which supports the schools], in England, the trade association, and in America, private enterprise which undertake this task."[25]

The Technical School for Carriage Draftsmen and Mechanics received support from the Carriage Builders' National Association until 1916, the end of the Carriage Era, when technical education was directed toward the automobile.

To examine the aesthetics of horse-drawn vehicles, especially from the perspective documented by those directly involved with their production, and the tastes established by elite consumers, requires an analysis of the various media and skills applied to the finished work–from the draftsman who created the total design to the trimmer who lined the interior with fabric–to illustrate that the trade was concerned with the creation of products both visually appealing and harmonious in the combination of their disparate parts.

Carriage fabrication required expertise in manipulating many different materials[26] and the complex composition of carriages lent itself readily to the subdivision of labor. The separate specialties required in carriage construction resulted in two major manufacturing formats, developed during the nineteenth century: vertically integrated factories where all phases of carriage production were executed and completed in a single establishment, and shops or factories that assembled parts made by accessories companies. Both assembly-line production and the vertically integrated factory employed a method of fabrication that corresponded to the discrete parts and materials applied to carriage construction.

In larger, metropolitan firms, the design of a specific type of vehicle was worked out in a series of drawings (see pages 21–23 for illustrations). The scale drawing suggested the overall shape of the vehicle in silhouette. The working drawing was more mechanically specific and geometrically precise, employing the elevated drafting technique of the "French rule," defined as

The application of practical geometry in the construction of carriage bodies, as practiced in France . . . . By this rule the body maker obtains the points by which he draws the correct side-sweep for the different pieces of the framework, the turn-under and side-swell being given by the operation of right lines drawn over the side elevation and cant of the body.[27]

Working drawings were transferred to a blackboard as a guide to the workers. A final drawing, made for the benefit of both worker and client, was rendered in pencil, ink and gum arabic. The colored draft indicated the finished appearance of a carriage, and required a "variety of touch, and that delicate handling of the colors employed, which at once stamps a colored draft as first-class in every particular, [and] depends not a little on artistic taste."[28] In coloring a vehicle design, a draftsman had to choose the color appropriate to the style of vehicle and the quality it should convey. In an example of a draft of a landau, the following advice was offered:

Colors give the appearance of lightness and heaviness, according as they are light and dark; and as we wish the Landau to appear rather heavy because it is large, taken as a whole, we decide that carmine or any light brilliant color would be out of keeping. Further the trimmings should agree with the color used on the panels, and if carmine is suitable for the panels, it or something approaching to it in color, is suitable for the trimming, thus carried out practically, we should have a bloody or fiery-looking piece of work, offensive to the person of good taste.[29]

The colored draft worked out the outline of the form and the harmony of the colors applied to the style of vehicle to portray the compositional harmony of color and form. This was a critical stage of carriage design. "No matter how well shaped the body, how tasty the carriage, how finely painted or elegantly trimmed the job may be, if there is no harmony in the different parts there will be no beauty in the whole."[30]

The contour and shape of a carriage relied on the combined mechanical skills of the body maker and the smith. Manipulation of wood, iron and steel gave a vehicle its strength to "bear the load imposed, as well as to give the shape of the several parts . . . to make up in general form what is known as a carriage."[31] The body worker had to understand the properties of the various woods used in body construction. Hardwoods were used to make structural and weight-bearing members; pliable softwoods were used to make body panels. Knowledge of the hygroscopic properties of wood and its proper seasoning were required to steam bend shafts or shape broad thin sheets of wood over the framing without splitting them. Carving, although not practiced widely by woodworkers in the carriage trade, was applied to sections such as pump handles, risers and blocks. Like other forms of ornament, carving was subordinate to the overall form.

The body hung on the gearing, also referred to as the *carriage* or *undercarriage*. The smiths and woodworkers who fabricated the gearing were responsible for the most mechanical part of the carriage. The springs, which absorbed shock and horse motion; the strength of the wheels; and the supporting members of the body, had to be equally harmonious with the composition and provide the means for the locomotion of the carriage. The maker of the gear was required to have a " 'good eye,' for patterns are but little use except to assist in getting sweeps, so the greater part must be done with the eye only as a guide; then again, he needs it in giving the carriage a proportion to correspond with the body, for each part should be in harmony with the other, and not look, when it is done, as though it was made of remnants from different kinds of jobs."[32] Thus both the wood and iron used to

provide strength and the mechanical parts and construction implemented the design. These elements established the shape and character of a carriage. To further enhance the artistry of a vehicle, paint and trimming were applied as "articles capable of supplying comfort, beauty and excellence [which were necessary] to produce a perfect whole."[33]

Carriage painting was a facet of the industry that was not influenced by industrialization. Robert Shinnie summarized the lack of progress in mechanizing that aspect of carriage manufacture in a paper presented to the Institute of British Carriage Manufacturers in 1896:

The system of carriage painting we follow today, in its entire process from priming to finish, is the same traditional one inherited and practiced by our fathers. Their aim was to produce a smooth and durable finish; our efforts are to accomplish the same.

This stationary condition of a department otherwise so changeable in taste and expression, sometimes suggests that dearth of inventive genius, and the absence of that progressive spirit that has fostered and developed the other branches of our trade to such perfection—while axles, springs,

Advertising Sample: "Senours Superfine Carriage Paint"
c. 1890
Senours Manufacturing Company, Chicago, Illinois
Purchase, 1981

Gear Colors

Can be furnished only as described under each job.

Yellow	Gray
London Smoke	Olive Green
Carmine	Brewster Green
Maroon	Blue
Carmine—Fancy Southern Striping	Black

head-work, and almost every moving part of a carriage shows inventive progress and bristles with patents, the painting remains a traditional sequence.[34]

Until the last quarter of the nineteenth century, the painter ground his own pigments with a mortar and pestle and mixed his own colors in various media. The introduction of paint mills expedited this process. Pre-mixed carriage paints in cans or tubes became available, but the painstaking method of application remained unaltered.

Carriage painting can be divided into two categories: compositional and ornamental. Compositional painting refers to the application of paint to different sections of noncommercial vehicles to integrate parts and to convey characteristics appropriate to a given style. Ornamental painting was applied to trade vehicles such as stage-coaches, omnibuses and business wagons.

These vehicles were intended to capture the attention of viewers and to display their goods, and were highly decorative. This branch of painting was defined as "a limit-less art, resourceful, restive, responsive to an admirable degree to the ever-varying side lights of technical skill. All that art can be anywhere the broad surface of the modern business wagon invitingly offers to display."[35]

The painter of private noncommercial vehicles had to choose colors that would complement and enhance the form, to unify or distinguish different parts as dictated by the overall design of the carriage. Color also had to be suitable to the use of the vehicle:

In choosing the color of a carriage, various considerations intervene:–whether it is to be used in summer or in winter, or in both seasons;–or whether it is intended to look rich in the outset, without regard to wear; or whether the chief consideration be durability, and the next appearance.[36]

(opposite)
Gear Colors
1913
From catalog No. 802,
Studebaker Brothers
Manufacturing Company
South Bend, Indiana

Summer Brougham
1901
Brewster & Company
New York, New York
Gift of Ward Melville, 1971

An example of imitation cane work can be seen on the lower quarter panels of this vehicle, built for Alfred Gwynn Vanderbilt in 1901.

Although a color scheme of a vehicle was often chosen by the draftsman or the customer, it was also determined by the painter. The purpose of color arrangements was to create a visually pleasing whole. If the running gear–the mechanical parts comprised of axles, reaches, wheels and springs–were cumbersome or obtrusive, it could be made to appear light by applying a contrasting color to "pick out" or stripe the members to create a "pleasing optical illusion."[37] "The chief and essential purpose of striping is to impart a beautifying effect to the surface upon which it is used."[38]

The painter received a vehicle "in the white," or unfinished. The surface had to be prepared to receive paint. Sanding and filling smoothed the wood and ironwork before the primer was applied to the interior and exterior. Primer consisted of pigment, oil, rubbing varnish and lead. Thinly applied to seal the grain, it acted as a preservative. A lead-based substance called "rough stuff" was laid on in a series of applications; once it was dry, extensive rubbings followed to assure a smooth surface. A "stain coat," a thin layer of paint, was applied to reveal inequalities in the surface which would be further rubbed out until smooth. A "color coat" of opaque pigment of high molecular density for good coverage was brushed on in a series of applications. Then transparent

From Sample Book of Transfer Ornaments
c. 1880
Published by Palm, Fechteler and Company New York, New York; Chicago, Illinois; and Montreal, Canada

34   PALM, FECHTELER & CO., NEW YORK CHICAGO MONTREAL

The decorations as shown on this sheet are made up specially for buggy decorations and are sold in sets. As many customers arrange the set to suit themselves, we prefer that you state how many pieces of each style you want, and how many sets you will buy, and we will then make you special net prices.

S 113 A
Gilt scroll flowers in red and white colorings

H 499 and S 113 A
Gilt scroll flowers in red and white colorings.

H 500
This design gilt scroll, red and green coloring. white and yellow high lights

H 502

H 501
Gold scroll with either red. green, brown or blue coloring

H 501 A
Gold Scroll or aluminum silver scroll

H 502
Ribbon in wine color. blue, green or brown
End pieces in colors to match ribbon scroll effect in silver or gold as wanted

pigments such as Indian yellow, Prussian green, carmine or alizarin crimson suspended in a varnish medium were applied in successive layers. The ratio of pigment to medium was reduced until the final "clear coat" of varnish was brushed on. Glazing coats were important to the optical effect of the color. Glazes also enriched panels painted black; tinting the varnish with black pigment diminished yellowing, which could make a black vehicle appear green. Richer, deeper pigments had little covering property. Light passed through the clear and glazed coats to the opaque color coat, imparting depth to the color; the glaze and varnish coats also "wetted up" the surface to give the vehicle a lustrous finish.[39]

Each consecutive layer was applied and rubbed by hand. Drying time depended on the weather and location of the shop as much as on the media and driers added to the pigment. The layout of the painting room and the quality of brushes and materials influenced the quality of the job. A vehicle had an abundance of irregular surfaces–rounded spokes, curved arch panels, moldings on the body and broad quarter panels–yet each section had to be finished to a high degree of excellence.

If deemed appropriate, striping was applied with striping or dagger pencils which were long, thin brushes made of camel, sable or ox hair. With these, the painter made a variety of striping patterns, varying from lines less than one-sixteenth of an inch thick to broad stripes that could be one inch thick. Patterns consisted of *hairline, fine line, stout line, round line, narrow line, heavy stripe* and *broad stripe*[40] or a combination of these. Striping required a steady hand and a superb eye.

Carriage painters also demonstrated an ability to imitate materials. Some vehicles, like furniture, were painted *faux bois* (false or imitation wood) to simulate or exaggerate wood grain. Another popular contrivance was imitation cane work; cane panels lightened the appearance of a carriage and added a certain amount of style. The procedure was quite complicated:

After the panel is finished, the laced or woven cane [used as a pattern] is inverted upon it with white chalk, and is afterwards filled up with a tint of pure blue cobalt, moistened and ground with essence of turpentine. The ground on which it is applied is, generally speaking, ultramarine blue, much deeper than the cobalt, and is afterwards shaded with smalt mixed with ultramarine which darkens and gives it the necessary body, without making it too thick. When the work is done, and well dried, the whole panel is repolished, and the last coat of varnish is given.[41]

Another form of ornament that became popular was the decal. These designs could be transferred onto a vehicle and were frequently used on sleighs or inexpensive carriages of midwestern manufacture. Transfer ornaments were comprised of black and white or colored lithographs "in exact imitation of the original paintings specially adapted to carriage work."[42] They included ornamental or figurative decorations and were published in catalogs.

Paintings that decorated trade vehicles were elaborate, serving as mobile advertisements for a company's wares. Unlike the painted and sculptural ornamentation applied to private, aristocratic carriages made prior to the nineteenth century, the ornamentation of trade vehicles did not obscure their form, except for novelty business wagons made in shapes that conformed to the product. Commercial and public carriages were decorated to attract attention; the trade vehicle was intended to deliver both the product and the stimulus to purchase it. The images on trade vehicles could be specific or evocative: Ice wagons frequently featured images of Eskimos, polar bears and other pictorial suggestions of cold and ice. A prime example of carriage painting and decorating for the purpose of attracting attention is set forth in a description of a medicine wagon built for the Cherokee Medicine Company of New York, which

has many novelties in construction, and is intended as an advertising medium. The front of the car behind the driving seat is raised in the form of a gallery, with swept sides embellished with carved tigers' heads, for the seating of a band, the seats of which are upholstered in claret leather. An operating chair with a high carved back extends across the middle of the car, and is richly upholstered in claret velvet, while the hind part of the car is seated round with sofa-shaped

Tea Wagon
c. 1910
Biehle Wagon and
Auto Body Works
Reading, Pennsylvania
Purchase, 1955

seats, beautifully carved and trimmed with claret leather. In front, on each side of the driving-seat are two statuettes, 3 feet high, one representing an Indian, and the other the figure of Fame. The sides are triangular and are ornamented with two very high-class oil paintings, one representing Prairie life and the other an Eastern scene, delineating the properties of medicines. The paintings are done by different artists, and give the car a remarkable and artistic appearance. In the sides and back are framed silver-plated mirrors with beveled edges, giving the names of the different medicines in embossed gold letters, shaded with red. The sides and back of the car are beautifully carved with buffaloes' and rams' heads, leaves, etc., and the wheels are fitted with carved face pieces, about 2 ft. in diameter. The body of the car is gilded with deep gold, while the underworks are painted a deep yellow, picked with gold, and relieved with red lines. The mountings are all brass. The whole has quite an artistic and finished appearance, and reflects great credit upon the builders, who have been highly commended for their work.[43]

Public conveyances such as hotel coaches and stagecoaches were also noted for the skillfully executed paintings that decorated their surfaces. The leading manufacturer of such conveyances was the Abbot-Downing Company of Concord, New Hampshire. Generalized specifications for designs were recorded in their Customer Order Books: "Ornament it up tip top style 7 good door orniments/Orniment neat but not Gaudy/Orniment nice Female Figure in dress & orniment up showy/Orniment up nice & Rich—no *Female* on *doors*.[44] The chief artisan in charge of ornamental painting employed by the Abbot-Downing Company was John Burgum (1826-1907). Burgum was born in Birmingham, England, and in 1850 moved to Boston, where he secured employment as a coach painter. That same year, he transferred to the Abbot-Downing Company, where he served as chief ornamenter until his death. In addition to painting coaches for the Abbot-Downing Company, he ornamented steam pumpers for the Amoskeag Company in nearby Manchester.[45] His work also included numerous sketchbooks and easel paintings.

Burgum was typical of a class of artists, many of whom were anonymous, whose

works varied from easel painting to decorative art. Because much of their work was unsigned and appeared on the surfaces of objects that fall outside the perimeter of "fine art," and because much of this art has worn away as a consequence of use, exposure or overly-zealous restoration processes, little is known of them or their work. Some carriage painters, by virtue of their ability to draw, became illustrators. Others were itinerant artists who also supported themselves by painting signs, carriages or furniture. Analysis of artists' work is contingent not only upon surviving works, which show the evolution of a style or development of technique, but also on

the study of preliminary sketches, notebooks and other records germane to the process of creating a work of art. Carriage painters regarded themselves as artists; existing examples of their work testify to the truthfulness of this. One such example is an omnibus, or barge, called the "Grace Darling" which was built by the Concord Carriage Builders of Concord, New Hampshire, about 1880. Although little information exists about the company, which occupied the New Hampshire prison works from 1876 to 1891, it has been assumed that it borrowed workers from the more established neighboring firm of Abbot-Downing. Similar in construction to the large

omnibuses built by the Abbot-Downing Company and the John Stephenson Company of New York, the "Grace Darling" is 23 feet long, 11 feet high and 8 feet wide. It was used by Simeon P. Huntress, who owned and operated a livery service in South Berwick, Maine, from 1860 to 1904. In addition to being a public conveyance, it was a mobile panel painting, extensively decorated with the work of John Burgum.

In spite of its immense size, "Grace Darling" appears delicate. Part of this impression is attributable to the compositional nature of the ground of the paintings. The band around the top, body and seat riser is painted a brownish red. The railing that comprises the backrest is straw-color, or naples yellow. Between the body and this rail are spindles, also yellow, with a black center; the undercarriage, like the railing, is straw-color and decorated with stout red and fine blue striping. The alternation of dark and light bands of color breaks up the mass of the vehicle.

The detail of the ornamental painting shows considerable artistry in execution. The letters on the side of the body, which spell "Grace Darling," are double block, with a gold-leaf face. They are shaded with glazed blue bands, implying dimensional recession. The gold of the letters is edged in a fine, opaque yellow line, which highlights the metallic surface of the gold leaf. The letters are framed by gold scrollwork, shaded and glazed in thin applications of browns accentuated by brilliant red centers. A one-inch gold-leaf band accentuated by a fine black line and shaded in a red glaze connects the scrolled borders.[46]

Each of the four corners is decorated with medallions that feature different representative paintings: a stag, a falcon and still lifes of fruit and flowers. The lighting in the stag and falcon paintings is luminous; the light emanating from the background of the stag highlights his ears, his beard and the edge of his back. The warm tones used to depict his hide contrast to the cool, purple tones of the sky in the background. Both the stag and falcon are more lively and naturalistic than the fruit and flowers painted in the still-life medallions, that are more formulaic. A separate still life painted by Burgum (n.d.) shows similarity in style to those on the "Grace Darling," with a more amateurish composition.

The seat risers are decorated with portraits of idealized female figures bordered by a shaded gold-leaf border and scrolled gold-leaf support. The figure on the left side is a mature woman in semi-profile; indirect light illuminates her features and accentuates the ruffled edging of her gown and the feathers that adorn her hair. The figure on

Still-life, Watermelon, Orange and Glass Compote
not dated
John Burgum (1826-1907)
Oil on canvas
Photograph Courtesy of New Hampshire Historical Society, Concord, New Hampshire

Omnibus,
"Grace Darling"
(details)

the right side, a younger woman also in semi-profile, is equally detailed.

The rear door summarizes the name of the owner through repeated associations. On the arch of the door is painted in red "S. P. Huntress," the owner's name. On the upper section of the door panels below the window, framed by ornamental scrollwork, is a figure of a woman in mid-nineteenth-century dress who is holding a bow and quiver, a depiction of the name "Huntress" as a nineteenth-century Artemis. Below the panel is a hound bringing down a stag, another reference to the hunt, executed in deft, confident brush strokes. Below this panel is the single-block shaded gold-leaf number "23."

The decorative surface of the vehicle varies from metallic, shaded gold-leaf borders and letters to the alternating reds and blues in the decorative medallions in the seat railing; the contrasting hues of the vehicle demonstrate a full range of the palette. All sections fit together in the composition of the body.

The interior was reserved for seated passengers. Between the pillars supporting the roof are small landscapes set in gray, Eastlake-style borders on an off-white background. Each scene is different, offering the traveler something to capture his or her attention while waiting to arrive at a given destination. All are views of nature not unlike those the passenger would encounter when looking out through the open spaces of the omnibus, which was used to convey people to scenic areas.

Both similar and near-identical images exist in a sketchbook illustrated by John Burgum in 1879. The dated sketches are drawn in an oval format, like the paintings, with additional notations recording the time of day and, in some instances, the location. These notations indicate that the sketches were drawn outdoors. The composition of both sketches and landscape paintings is asymmetrical, with the compilation of forms and weight on one side. Pencil marks and brush strokes depicting the scenes are equally short and direct. It can be surmised from comparing the sketches to the paintings that the artist made freehand drawings for his own pleasure and used these images in paintings commissioned for vehicles at a later time.

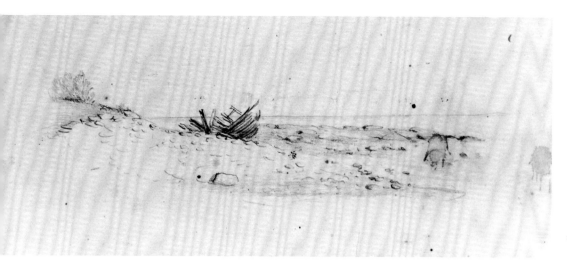

Sketch of a Shipwreck
not dated
John Burgum (1826-1907)
Pencil on paper
Photograph Courtesy of
New Hampshire
Historical Society,
Concord, New
Hampshire

**Omnibus,
"Grace Darling"
*(detail)***

By contrast, the flowers painted on the corner of the body and the door interior are more stylized, implying tracing of a standard design. In this technique, a pattern of flowers that was heavily outlined was pin-pricked for transference of the design onto the vehicle. This type of ornamentation was more common than direct painting. Original designs were either worked out by the artist or borrowed from published manuals or trade journals and traced on the surface of a vehicle. Regardless of their source, these designs were painted in great detail, with glazing, accentuating lines and highlights–elements that were barely discernible when a vehicle was in motion. Both ornamental and compositional painting were executed in refined and labor-intensive procedures to contribute to the artistry of a vehicle.

Trimming was the branch of carriage manufacture that most directly appealed to the comfort of the passenger. Enclosed vehicles such as coupés, broughams, coaches and more formal versions of the rockaway featured personal and intimate interiors. The somber exterior of coachman-driven vehicles established a barrier between the occupant and the public. Closed carriages served both as conveyances and as mobile shelters; their interiors were quiet. The fit of doors and windows was exact; shades could be drawn over thick, beveled glass windows. Inside, the environment was comfortable and exclusive. Interior furnishings in tooled leather, dark mahogany or carved ivory held the personal effects of the passengers.

Both the quilted satin that accommodated passengers in formal carriages and the leather folding top of a humble buggy were products of the trimmer's art. As with other skills applied to carriage manufacture, "harmony is at all times an essential element."[47] The trimmer's choice of materials and fittings, often specified by the customer, was subordinate to the color, shape and style of the vehicle.

A trimmer needs to be an artist; he must be an

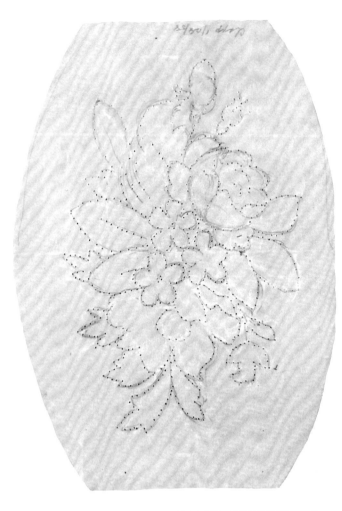

expert in colors, so that the style of trimming selected may produce the best results. Such an one would select laces and all interior mountings that would harmonize with the material used, and the pattern selected. He would be governed by the form of the body and the colors used in painting, and by a harmonious combination and artistic construction he would secure the desired result.[48]

The trimmer was responsible for cutting, stitching and installing dashes, side curtains, folding and extension tops, fenders and other, more functional parts of a carriage. It was important for the trimmer to know the properties of his materials; fabrics and leather used in trimming were the most susceptible to deterioration. Moths, dust, fugitive dyes and the adverse effects of sunlight on fabric fibers and leather finishes were potential hazards.

While the carriage trimmer is interested mainly in his special branch, he should know how to care for the complete carriage, for whatever injures any one part of it injures another, and no one portion is more easily soiled than the trimming. Fine cloths, laces, silks, leather and mountings, all require care and attention to preserve them in such a manner as will insure their durability and beauty.[49]

Careful preparation and expert manipulation of materials were necessary to delay deterioration, including that caused by natural weaknesses. The grain of the leather used for roll-up curtains conformed to the direction imposed by rolling. When using leather, the trimmer had to consider the type of hide, the tanning process and the section of the hide that would best suit a given purpose. Fabrics were chosen for their durability and suitability; more fragile fabrics such as damask and satin were reserved for the protected and private interiors of enclosed vehicles.

Materials used in interior trimming included morocco (goatskin), wool, satin and damask. These were pleated to make window and seat valances, or quilted and padded with curled horsehair (preferred for durability), moss or excelsior. Lifters, holders and wide, ornamental borders were made of broadlace, which was made in two-and-a-half-inch strips and could be rep, with a plain figure, or cut, for a patterned figure.

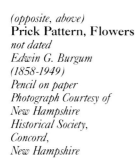

(opposite, above)
Prick Pattern, Flowers
not dated
Edwin G. Burgum
(1858-1949)
Pencil on paper
Photograph Courtesy of
New Hampshire
Historical Society,
Concord,
New Hampshire

(opposite, below)
Omnibus,
"Grace Darling"
(detail)

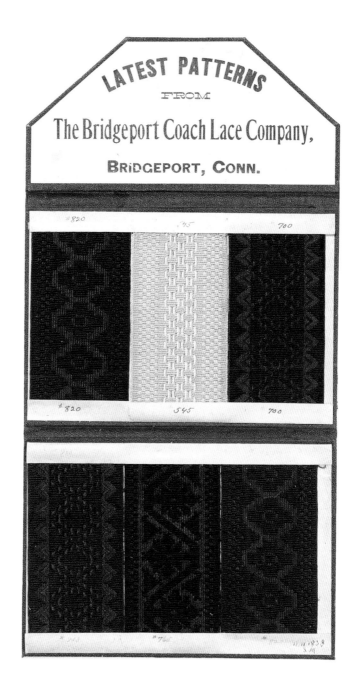

Lace Sample,
"Latest Patterns"
c. 1880
The Bridgeport Coach
Lace Company
Bridgeport, Connecticut
Purchase, 1955

The fiber content of broadlace was generally Scotch linen for the ground warp, cotton thread for the slack ground warp and worsted silk for the face, running parallel to the warp. Filling was made of cotton or linen.[50]

Although some laces were made by hand, most were designed and manufactured by specialty companies and made on power looms. Seaming lace, a narrow fabric with one selvage, was used for covering cords and joining seams; pasting lace, a narrow fabric with two selvage edges, was used for borders and finishing edges.

Additional trimming materials included tassels, fringe, tufts and fabric-covered buttons. Fittings for the interior consisted of card and cigar boxes; vanities; fan, cane and parasol holders; and call bells for summoning the driver. Silk curtains or roll-up shades, which could be rolled up automatically by engaging a spring-loaded trigger, were also used. All of these had to be combined to create a harmonious and comfortable interior.

A trimmer, like a painter, was advised to develop a knowledge of color and color harmonies to maximize the total effect of interior trim.

A study of colors, the effects produced by certain shades when given certain forms, harmony of the whole when grouped, inside ornaments and artistic designs, are all absolutely necessary if the workman aspires to be more than a mere automaton.[51]

The trimmer who is able to select colors and designs which harmonize with the form and character of the vehicle, and to give the touch of an artist to all minor details, producing a luxurious and comfortable lining for a carriage, may claim a high position as an artist.[52]

This artistry is evident in the interiors of two different carriages in The Museums collection, a family coach and an enclosed sleigh. The coach, made for the Gardiner family of New York by James Brewster and Sons about 1848, is trimmed in off-white wool. The heavily padded backrest provided a comfortable support for the passengers. The squabs, or upholstered interior sections of the quarter panels and the lower door

panels, are diamond-pleated. The figure of the broadlace is off-white, set off against a light, white silk background. The net-headed speaking tube and the lifters and shade pulls are finished with a bullion fringe, a type of fringe made of heavy twisted cord. The drop handles, slides and check string pulleys are ivory, harmonizing with the off-white interior.

The interior of the chariot-body sleigh, or "booby hut," made by J. T. Smith of Boston about 1880 evokes a very different atmosphere. This vehicle was used in an urban area during the harsh winter months. Its interior trim is a dark red plush velvet. The broadlace, silk shades and carpet are red. The stable shutters are made of a golden-blond wood in a dark wood frame. The ceiling is lined with pleated red damask. Unlike the interior of the coach, which has a pleasant and sociable atmosphere suitable to a family carriage, the interior of the sleigh suggests warmth as well as comfort and intimacy.

Most laces and fringe were ordered to match the fabric used in trimming. Although colors were usually consistent, combinations

(opposite, left)
Coach (interior)
c. 1848
James Brewster and
Sons, New York, New York
Gift of The Brooklyn
Museum, Brooklyn, New
York, 1961

(opposite, right)
Booby Hut (interior)
c. 1880
J. T. Smith
Boston, Massachusetts
Gift of The Society for
the Preservation of
New England
Antiquities, Boston,
Massachusetts, 1951

Circular-front Coupé
(detail of interior
door trim)
c. 1860
J. B. Brewster &
Company, New York, New
York, Gift of Ward Melville,
1971

of colors were also employed. The use of different colors in trimming materials was not to create alarming contrasts; such an effect would have been objectionable. The satin-trimmed interior of The Museums circular front coupé made by J. B. Brewster & Company of New York about 1860 illustrates an attractive use of green and rose in the fringes, laces and rosettes.

Trimming materials, like vehicular styles, were subject to fashion. By the 1880s, with a few custom-ordered exceptions, satin became unpopular as a trimming material, and the more durable morocco was preferred. Silk was considered by some to be gaudy and fragile, and was also deemed potentially inharmonious with the dress of female occupants.[53]

Even the best becomes bad when overdone. The most elegant article is of bad taste when its refinement in material or workmanship makes it unsuitable to its purpose, unavailable for use. Ladies are generally very well aware of this rule with regard to their dress. A woman of taste will not wear a dress in the streets which cannot be worn without being soiled and spoiled. But with regard to the trimming of their carriages they seem liable to overlook the rule.

We have seen many carriages trimmed with silk. At one time silk seemed to be the most fashionable trimming for coupés and other closed carriages of first class. Crimson silk with black buttons and black laces was often used; but we remember, especially, one coupé trimmed with black silk and yellow buttons and laces. It looked very elegant indeed, but it looked a little too elegant. It had something so soft and tender about it, that it impressed us as if the coupé were made for a sick person, and when richer and gayer colors are used—for instance, crimson—the silk has a gloss which makes the appearance a little gaudy. Not that silk is gaudy by itself, but used for this purpose, the trimming of a carriage, it becomes gaudy. We know that one drop of rain or one atom of dust will spoil it, and we cannot help thinking that even the most elegant carriage ought to be able to withstand a little cloud of dust and half a thunder-storm without looking soiled.

But what especially strikes us as wrong when we see a carriage trimmed in silk, is the unfavorable contrast which silk forms to the dress of the lady who occupies the carriage. The interior of a carriage, like the walls of a room, ought to be of a modest, subdued, and quiet appearance, so that the human moving among these surroundings makes the splendor, the actual point of attraction. They form the background. They ought to be retreating. But silk, especially crimson silk, has a gloss and brilliancy of color which attract the eye in a very high degree, and make all other stuffs, even the most costly, fade away into insignificance; it makes a genuine cashmere shawl look like a rag.[54]

The trimmer had to be apprised of the guidelines of fashion, the suitability of styles of trimming to a particular vehicle style and the intended function of a carriage. The trimmer also had to know the properties of the various materials used to give dimension and artistry to the finished product. Each vehicle trimmed required a considerable range of materials and supplies.[55] Whatever the style, the trimmer was required to fit together all components with skill and harmony to accommodate the physical comfort and tastes of a customer and to complement the appearance of the vehicle as a whole.

The manufacture of carriages and the artistry applied to their construction, painting and finish were intended to supply a buying public with vehicles to suit their purposes—sport, pleasure, work or general transportation. The appearance of a vehicle and its components were dictated by taste—what would appeal to customers at a given time. Members of the industry contrived formulas and definitions to establish what shapes, elements, colors and materials were the most beautiful and the most desirable to employ,[56] but the applications of these tenets were influenced by the desire for variety and the fluctuating preferences expressed by public demand.

The aesthetics of carriages, however, did not exist solely to enhance sales and increase the profits of the industry. Carriages were intended for use, and in that capacity were mobile and available for all to observe.

But after all, those who do not use pleasure carriages—as the mass of the community, still benefit indirectly by them. They are objects of beauty constantly exposed to public view, as much so as architectural erections—perhaps more so, being locomotive,—and far more so, though in a

Central Park, New York City— The Grand Drive
December 16, 1865
A. R. Waud
Harper's Weekly

humbler way, than paintings or statues. It is an undeniable fact, that the daily habit of beholding beautiful objects has an imperceptible effect in refining national taste.[57]

Carriages were drawn by horses in harness, and the success of conveying the beauty of the turnout, or equipage, depended on the harmony not only of the passing silhouette of the vehicle, but also of its proportional relationship to the horse, passengers, coachman, groom and the carriage's appointments, which included a variety of furnishings and accoutrements, such as shooting baskets, lamps and vanities.

Carriages were indeed a "gift to the street."[58] The colorful trade vehicles and painted omnibuses supplied the riding and pedestrian public with mobile, albeit vernacular, art forms. The elegant turnouts that graced the urban parks in the nineteenth century were made to be seen as well as used. The variety of vehicles that paraded through city streets offered shapes, colors and movement to the urban environment. The splendid horses in gleaming brass or silver-plated harness pulling understated

broughams, the ladies in fashionable costumes driving their phaetons and the elegant gentlemen whips displaying their coaches and drags were fantastic spectacles for those who witnessed the daily cavalcade of vehicles. Lightweight road wagons pulled by fast trotters shared the roadway with heavy market new wagons and stately coaches. The most exclusive vehicles, manifestations of elite taste and purchasing power, became democratic on the streets, and were visible testimonies of the artistry of those who had created them. "In whatever position it is drawn, at whatever speed it may be going, a carriage, like a vessel, derives from the fact of motion, a mysterious and complex gracefulness which is very difficult to note down in shorthand."[59]

NOTES

1 Henry Julian, "Art Applied to Coachbuilding," 29 April 1884, *Papers Read before the Institute of British Carriage Manufacturers, 1883-1901* (Aspley Guise: The Powager Press, Ltd., 1902) 57.

2 France became a center for artistic production during the reign of Louis XIV in the seventeenth century. The King's Superintendent of Finances, Nicolas Fouquet, initiated the centralization of all decorative arts (including carriages), which, under his strict supervision, helped to create a national style. This system continued through the eighteenth century, when it was disbanded by the French Revolution.

3 Ralph Straus, *Carriages and Coaches* (London: Martin Secker, 1912) 204.

4 George W. W. Houghton, "The Coaches of Colonial New York; a paper read on the evening of March 4, 1890 before The New-York Historical Society," (1890) 31.

5 Furniture, boats and carriages used similar materials and technology, and steam bending was important to each of these industries: "Few know that the fine carriages they ride in are very largely made of bent woods. The felloes of their wheels are bent, and are made in two parts. The framework of coaches and heavy carriages is nearly all made of bent stock. They are not only better made but are more cheaply made. The frames of most of our pleasure boats are bent, and so are many of the frames of some of our finest sailing yachts. Furniture of many kinds has bent frames. All the celebrated Thonet chairs, which for comfort and beauty are not excelled in the world, are entirely of bent wood. The object of bending is two-fold–saving of time and stock, and stability and strength of the work when put together. We ought to add another; beauty of form" (*The Hub*, 31.2 1889: 197.)

6 Ralph Nicholson Wornum, "The Exhibition as a Lesson in Taste," *The Crystal Palace Illustrated Catalogue*, originally published by George Virtue, *The Art Journal*, 1851 [New York: Dover Publications, 1970] lxxx.

7 "America produces hickory and other woods tougher and lighter than any grown in other countries; and of these American mechanics produce marvelous specimens of ingenuity in strength and lightness, in two-wheeled and four-wheeled vehicles" (S. Sidney, *The Book of the Horse* [London: Cassell, Petter, Galpin & Co., 2d ed., n.d.] 386.)

8 *The New York Coach-Makers's Magazine*, 2.10 (1860): 197.

9 *The Hub*, 31.8 (1889): 585.

10 "Avoiding details, it would appear that in the design, construction, colors, and appointments of carriages, certain fundamental rules may be laid down.

"In every case, fitness, or adaptability of the carriage to the purpose for which it is required, should be the first consideration; without this, whatever may be its beauties of form or harmony of colors, it must result in a failure.

"Simplicity of construction, with thorough effectiveness and quiet elegance of outline, is the next great point.

"Individuality of character is most important, not only in the *tout ensemble*, but in all matters of detail: '*Quisquid in suo genere satis effectum est valet.*' " Due proportions, free from exaggeration, and independent of any passing style and mannerism, are most essential. The employment of material, however plain and simple, the best of its kind, carefully selected, and adapted by its peculiar quality to the particular requirements of the case, this, as a matter of course, must be united to excellence of workmanship.

Except in carriages built expressly for display, *a subdued style of painting* is most permanently pleasing, and the avoidance of anything *outré* in all decorations and appointments should be studied; thus forming a certain harmonious combination

which, without attracting attention to any one part, pleases and gives satisfaction as a whole." (*The New York Coach-Maker's Magazine*, 2.8 1860: 145).

11 Julian 57.

12 *The New York Coach-Maker's Magazine*, 2.10 (1860): 196.

13 George Fleming Budd, "Coach Body Making," in *Reports on Carriages in the Paris Exhibition* (London: Hardwicke & Boque, 1879) 26.

14 *Harper's Weekly*, 21.1047 (1877): 46.

15 James Garland, *The Private Stable* (Boston: Little, Brown and Company, 1903) 63.

16 Garland 74.

17 Garland 74-75.

18 Henry William Herbert, *Hints to Horse-Keepers* (New York: A. O. Moore & Co., 1859) 374-375.

19 *The Hub*, 31.8 (1889): 589.

20 *The Hub*, 31.11 (1890): 854.

21 Eleventh Annual Report of The Metropolitan Museum of Art, May 1, 1881, 196. The Metropolitan Museum of Art Archives.

22 *Prospectus: Class in Carriage Drafting and Construction*, New York, December 21, 1881. New Hampshire Historical Society, Archives and Manuscript Division.

23 Letter of Announcement. New Hampshire Historical Society, Archives and Manuscript Division.

24 Sixteenth Annual Report, The Metropolitan Museum of Art, December 31, 1885, 318. The Metropolitan Museum of Art Archives.

25 *The Hub*, 19.4: 610.

26 "Carriage materials, comprise a great variety of articles manufactured from many different substances, out of the number of which we mention iron, steel, brass, tin, lead, gold, silver, oride, wood, leather, silk, wool, cotton, hemp, minerals, seeds, gums, ivory, bone, etc. etc. The finished carriage is required to possess strength, durability, comfort and beauty" (*The Coach-Makers' Illustrated Hand-Book*, 362).

27 Don Berkebile, *Carriage Terminology: An Historical Dictionary* (Washington, D.C.: Smithsonian Institution Press and Liberty Cap Books, 1978) 345.

28 *The Coach-Makers' Illustrated Hand-Book* 40.

29 *The Coach-Makers' Illustrated Hand-Book* 40–41.

30 *The Coach-Makers' Illustrated Hand-Book* 95.

31 *The Coach-Makers' Illustrated Hand-Book* 362.

32 *The Coach-Makers' Illustrated Hand-Book* 95.

33 *The Coach-Makers' Illustrated Hand-Book* 362.

34 Robert Shinnie, "Carriage Painting," June 1896, *Papers Read Before the Institute of British Carriage Manufacturers, 1883-1901*, 476.

35 M. C. Hillick, *Practical Carriage and Wagon Painting* (Chicago: Press of the Western Painter, 1903) 109.

36 William Bridges Adams, *English Pleasure Carriages* (London: Charles Knight & Co., 1837) 210.

37 Adams 212.

38 Hillick 61.

39 Although techniques varied, the standard process and

materials used were described in the "Paint Shop" section of *The Hub:* "Sand the body 'wherever the grain shows' and dust; prime with Venetian red, ground fine in oil, mixed with lampblack in a medium of 2/3 calcutta raw linseed oil and 1/3 high grade rubbing varnish (this mixture must be strained) – mix in turpentine just before applying; apply primer to the inside first and then the outside to seal the pore of the wood and to protect it from damage. Allow to dry for 5-6 days. Apply 'roughstuff' made with three parts English filling, one part keg lead, mix with good quality rubbing varnish and run through the paint mill. Cover until needed for use. Once applied, rub with pumice. Apply stain coat, pigment mixed with turpentine and rubbing varnish. When stain is rubbed down, wash the body and wipe with chamois. Allow 5-6 hours to dry (for moisture to evaporate). Apply a coat of PWF (permanent wood filler), rubbed in well to seal pores. Test the color, working paint down in horizontal strokes; do not reapply brush to areas already painted. Apply a series of color coats. Mix clear color in varnish and apply several coats. Once dry, polish with stone (pumice or rotten) and felt. When the painting is complete, place the vehicle in a dark room" (*The Hub*, 31.11 1890: 846–847).

40 Hillick 63.

41 *The New York Coach-Maker's Magazine*, 3.10 (1861): 191.

42 *The Hub*, 22.12 (1881): 595.

43 *The Hub*, 31.4 (1885): 268.

44 Abbot-Downing Company: Order Books, vols. 4-6, 1858-1902. New Hampshire Historical Society, Archives and Manuscript Division.

45 *Burgum Family Papers*, compiled by Thomas E. Camden, March 1982. New Hampshire Historical Society, Archives and Manuscript Division.

46 An example of the complexity of painting ornamental borders is described in an article in *The Hub* entitled "Renaissance Corner Pieces": "In the first place trace out a good drawing and prick same with pin holes; this is called the ponce pattern. Next get some good stencil paper, or foolscap will answer, oil the paper with one part oil and two parts japan, wipe off dry, lay ponce pattern on and ponce stencil paper with ponce bag, made of ground charred coal or a little Prussian blue and whiting. After poncing, proceed to cut the stencil. Cut out the places at the border that show dark, leaving the light which is supposed to be gold or aluminum leaf. If the latter, it will have to be glazed with lacquer to represent gold. The decoration being in gold leaf and the ground color in panel vermillion, stencil the band with a color a few shades darker than the ground color, one part Tuscan or Indian red to six parts vermillion. The leaf scroll, which forms the basis for the corner piece, should be etched or shaded with fine black lines as in drawing, or if desirable, it can be glazed with asphaltum at the places that represent the dark, and to represent high lights or raised parts, glaze with yellow lake, which will make a very pretty effect. Treat the top corner rosette the same way. If desired, the ribbon parts could be glazed with ultramarine on the vermillion ground. Edge the whole with a fine line of black between stripes 1 and 2. Glaze with carmine" (*The Hub*, 31.1 (1889): 42).

47 William N. Fitz-Gerald, *The Carriage Trimmers Manual and Guide Book* (New York: NP, 1881) 21.

48 Fitz-Gerald 22.

49 Fitz-Gerald 49.

50 Fitz-Gerald 46-47.

51 Fitz-Gerald 164.

52 Fitz-Gerald 170.

53 Fabrics and colors were intended to harmonize with the total composition of the vehicle and also the occupant. An amusing quideline included in Henry William Herbert's *Hints to Horse-Keepers* stated, "Blue, in trimming, harmonizes with the complexion of a blond while it impairs the beauty of a *brunette;* with a crimson dress or shawl, blue is decidedly inharmonious" (Herbert 394).

54 *The Hub*, 14.10 (1873): 239.

55 An example of a list of trimming materials for a coupé includes "26 feet jap. trimming; 2 1/2 feet skirting leather; 5 feet hard splits; 3 feet soft splits; 6 feet railing; 14 1/2 feet grain dash; 9 goatskins, or 3 yards satin; 7 1/2 yards cotton cloth; 6 1/2 yards cotton drilling; 5 yards cambric; 1 yard Ewam; 60 yards narrow lace; 2 yards oil carpet; 2 3/4 yards Wilton carpet; 2 yards braid; 1 gong bell; 1 whip socket; 1 pair shaft tips; 1 pair shaft straps; 1 pair apron straps; 1 pair pole straps; 18 lbs. curled hair; 1 lb. wadding; 171 tufts; 21 buttons; 1 gross nails, covered; 1/2 gross screws, assorted; 2 knobs; 1 1/2 lbs. tacks; 3/4 lb. cord and twine; 1 oz. thread; 7 acorn tassels; 4 curtain tassels; 4 spring panels; 5 feet molding; 3 1/4 yards silk; 8 yards silk cord; 1 looking glass; 2 bent beveled glass; 2 French plate glass; 1 foot galvanized wire netting; 1 check cord; 4 lbs. paste; 2 each buckles and billets; 1 pair silver buckles; 3 bone slides; 3 yards webbing; 22 springs; 1 back light, beveled; 1 lb. risers; 1 set cushion straps; 1 pair frogs, silk; 1 pair frogs, ivory; 1 card case; 1 pair lever handles; 1 pair pull handles; 1 pair door handles" (Fitz-Gerald 117).

56 Some of these formulas were based on traditional aesthetic theories; others were pseudo-scientific, such as the explanation published in *The Hub*, "Why the Curve is the Line of Beauty." This article explained that the curve engaged more muscles in the eye, whereas rectilinear shapes used only two eye muscles, creating "a notion of tedium . . . a distaste for straight lines" (*The Hub*, 14.11 (1875): 255).

57 Adams 9.

58 This term is borrowed from a book about Victorian architecture, *A Gift to the Street* (San Francisco: Antelope Island Press, 1976).

59 "The Painter of Modern Life," *Beaudelaire: Selected Writings on Art and Artists*, translated with an Introduction by P. E. Charvet (Middlesex and Baltimore: Penguin Books Ltd., 1972) 434.

A RIDE INTO HISTORY:

The Horse-Drawn Vehicle in Selected New York State Counties, 1800 to 1920

Doris Halowitch

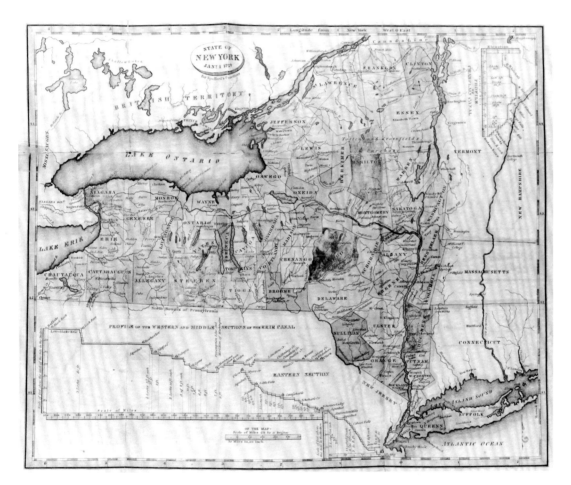

Map of the State of New York
January 1, 1824
From A Gazetteer of the State of New York
Horatio Gates Spafford
B. D. Packard,
Albany, New York

In 1800, Ontario County included the present-day counties of Genessee, Monroe, Wayne, Livingsto and Yates.

During the course of the nineteenth century, industrialization, urbanization, growth of population and increasing wealth transformed New York State demographically and economically. At the beginning of the century, New York State ranked third in the nation with a population of 589,000. The vast majority of its inhabitants, 514,000, resided in rural areas and were engaged in agricultural pursuits. By the end of the century, New York State ranked first among the states with a population of 7,269,000, of whom 5,298,000 were city dwellers and 1,971,000 were rural. It was also the nation's leader in manufacturing, with 78,658 manufacturing establishments, $1,651,210,000 in capital and $2,175,726,900 in value of goods produced.[1] The development of a transportation network, including horse-drawn transportation, expedited the transformation of New York State.

The manuscript copies of the tax records for New York State for 1799, 1800 and the Civil War years provide the statistical data base for this essay. (Details of the tax laws and tax records can be found in Appendix A.) These records disclose the number and the geographic distribution of privately owned horse-drawn vehicles reported for personal use rather than for earning a living. The 1799 and 1800 records list specific vehicles and their value; unfortunately, however, the owners of the vehicles are not identified. The Civil War records, on the other hand, contain the names of the owners but rarely designate the specific type of vehicle. (Appendix C contains descriptions of the vehicles mentioned in this essay.)

Three time periods are covered: 1800-1840, 1850-1870 and 1870-1920. In each period the focus is on areas of New York State selected as representative of diverse regions and different levels of economic development:

1 A highly commercial, urban, highly populated area: New York County and environs;
2 A rapidly developing economic crossroads: Albany County;
3 Rural areas, primarily agricultural regions: Clinton and Ontario Counties.

Several trends, which will be explored in greater detail, became apparent through analysis of the data:

1 Ownership of private vehicles remained the privilege of the wealthy during the nineteenth century;
2 Economic development of a region was a key factor in the extent of private vehicle ownership;
3 Few women are identified as owning vehicles in their own right, even though New York State's Married Woman's Property Act of 1848 and 1860 enabled married women to possess real and personal property in their own names;[2] and
4 There were definite differences in the pattern of ownership and use between urban and rural areas. In highly urban areas like New York County, primarily only the wealthy could afford to own and maintain private vehicles; a variety of public forms of transportation served the many.

In suburban or prosperous agricultural areas, per capita vehicle ownership was higher than in urban areas, because it was easier and less expensive to stable and maintain horses and vehicles in the countryside and because public transportation was less readily available. In these areas, such as parts of New York, Richmond, Westchester and Albany Counties, residents tended to have more expensive vehicles than those of less-developed, more rural areas.

THE EARLY YEARS, 1800-1840
Introduction: New York State in 1800

The transportation revolution in antebellum America is the subject of numerous histories. These studies, however, tend to underestimate the horse-drawn vehicle as a viable mode of transportation and to gloss over the critical contribution of horse-drawn vehicles to the development of a transportation and commercial network. To remedy this oversight with regard to New York State, it is necessary, first, to document the number and geographic distribution of specific types of privately owned horse-drawn vehicles in New York State in 1800 and, second, to highlight the role of horse-drawn conveyances in the state's evolving transportation system in the period from 1800 to 1840.

In 1799 and 1800 the New York State legislature levied taxes on real and personal property. Table 1 (pages 102–103) contains information culled from the Abstracts of Valuation compiled in 1800 for the taxes levied.[3] The 30 items of personal property specifically enumerated at predetermined values for taxation included

the eight types of horse-drawn vehicles listed in Table 1. The legislature's reference to particular types of vehicles reveals their familiarity with the appropriate nomenclature and their confidence that both taxpayers and tax assessors were similarly knowledgeable.[4] According to the abstracts for 25 of New York State's 30 counties, there were 41 coaches, 73 chariots, 13 post chaises, 226 phaetons or coachees, 193 other four-wheel pleasure carriages, 348 two-wheel top carriages and 1,600 other two-wheel pleasure carriages, for a total of 2,494. Not taxed were such horse-drawn vehicles as sleighs, stagecoaches, coaches used for hire and carts or wagons used to transport goods and refuse. Abstracts are not available for Dutchess, Oneida, Otsego, Saratoga and Steuben counties; the abstracts for 1799 list 279 ennumerated vehicles in those counties.

In mapping the distribution of carriages by county, a geographic pattern emerges. There were more enumerated vehicles in New York County than in any other county,

despite the fact that in 1800 New York County encompassed only the seven wards at the southern tip of Manhattan Island and the uptown Harlem division. (The wards were political divisions of the city for representation and local election purposes.) In the counties immediately surrounding New York County, there were 251 enumerated vehicles in Kings, 429 in Queens (Queens County in 1800 encompassed the present-day counties of Queens and Nassau), 258 in Suffolk, 118 in Richmond and 252 in Westchester. Thus 1,973 of the 2,494 total of enumerated vehicles were concentrated in New York County and adjacent counties. The farther north or west a county was from the New York City area, the fewer vehicles and fewer types of vehicles were to be found in it.[5]

New York County And Environs: A Commercial Center

New York County (Manhattan Island) was already a commercial center by 1800, with ferry connections to its neighboring counties and stagecoach lines to Boston and Philadelphia. It was also an important port for trans-Atlantic and eastern coastal shipping. The figures in Table 1 attest to the concentration of wealth in this county: The value of real estate far exceeds that of the other counties, as does the combined total of enumerated and nonenumerated personal property. The 665 vehicles listed encompass all categories of taxable vehicles. In the entire state, a total of 353 vehicles were in the $300-to-$800 category. Of these, 218, or 62%, were owned by residents of New York County: specifically, 117 of the 226 coachees or phaetons, 9 of the 13 post chaises, 61 of the 73 chariots and 31 of the 41 coaches. Of the total number of vehicles reported for New York County, 32.8% were in the $300-to-$800 category. Nevertheless, the most commonly reported vehicle, the two-wheel pleasure carriage (289, or 43.5% of the total of 665), was the least expensive. Despite the presence of a considerable number and variety of vehicles in this county, vehicle ownership was only 1.1%, and the affluent probably owned most of these vehicles.

The 60,489 residents of New York County were densely packed into the southern tip of Manhattan Island. Although the city limits gradually extended northward, the density of the population continued to grow.

By 1825 the population was 166,086. As the streets became increasingly congested, difficulty in getting from one place to another escalated. To alleviate this problem, a new mode of horse-drawn transportation was introduced by Abraham Brower in 1829: the sociable, forerunner of the omnibus. John Stephenson of New York City built the finest omnibuses, which initially carried 12 to 14 passengers. By 1835 there were approximately 100 omnibuses in New York. Whether these vehicles alleviated or aggravated the congestion is a moot point, however; one vivid 1837 account demonstrates that then, as now, pedestrians took their lives in their hands when crossing the streets of New York City.

Broadway is a noble street, 80 feet wide and straight as an arrow. . . . But . . . it is now quite too narrow for the immense travel, business, and locomotion of various kinds, of which it is the constant scene. This is particularly the case . . . below Canal Street and . . . south of the Park. Here the attempt at crossing is almost as much as your life is worth. To perform this feat with any degree of safety, you must first button your coat tight about you, see that your shoes are secure at the heels, settle your hat firmly on your head, look up street and down street, at the self-same moment, to see what carts and carriages are upon you, and then run for your life. We daily see persons waiting . . . for some minutes, before they

Omnibus Tickets
1830-1850
New York, New York
Gift of Mr. and
Mrs. William Crawford,
1977

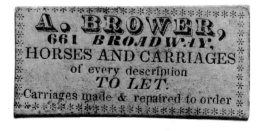

can find an opening . . . between the omnibuses, coaches and other vehicles, that are constantly dashing up and down the street; and, after waiting thus long, deem themselves exceedingly fortunate if they can get over with sound bones and a whole skin.[6]

Pedestrians, vehicles and refuse clogged the streets of New York; the cost of cleaning the streets during the period 1836 to 1838 was $355,901.75 after the reduction of an average of $45,594.55 per year from the sale of manure.[7]

Richmond County, perhaps better known as Staten Island, is an island of 58.5 square miles located in New York Harbor between New York and New Jersey. There were fewer enumerated vehicles in Richmond than in any other county near New York City, and the numbers are misleading until the size of Richmond's population is taken into consideration. In 1800 that population was 4,563, resulting in vehicle ownership of 2.6% or more than double the figure for New York State as a whole. Although the residents of this county owned more vehicles on a per capita basis, only three of the reported vehices are in the $300-to-$800 category. Of the total number of vehicles reported for this county, 97.5% were in the lower-valued categories and 75% were in the lowest-valued category.

Private means of transportation were augmented by ferry service to New York City and to Perth Amboy, New Jersey. A cross-county stage line was in service prior to the American Revolution. By combining the stage line and the ferry service, the inhabitants of Richmond were able to journey directly to Philadelphia. This service was advertised in the following manner:

Any Gentlemen or Ladies that want to go to Philadelphia in the stage and be home in 5 days and be 2 nights and 1 day in Philadelphia to do business, or to see the market days.[8]

Such travel was difficult: roads were poor and ferry crossings were hazardous, and the trip required several changes from stage to ferry and back again to a stage.

Westchester County, encompassing 525 square miles, had 21 towns and a population of approximately 27,428 in 1800. Bordered by the Hudson and East Rivers, it was blessed with many small streams that made it a fertile area. As is evident from the figures in Table 1, Westchester County was a prosperous area, and it is not surprising to find that the inhabitants owned a sizable number of enumerated vehicles of all types. Of the 252 vehicles reported for this county, 10.3% were in the $300-to-$800 category. Once again, however, the most commonly

Stagewagon Leaving a Tavern or Inn
1795-1820
Artist, F. Fumagalli (signed)
Hand-colored engraving
Philadelphia, Pennsylvania

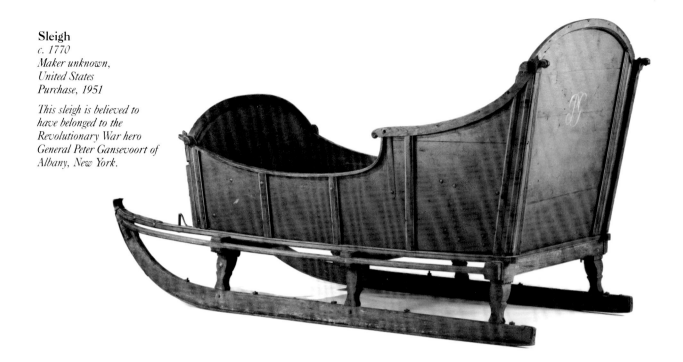

Sleigh
c. 1770
Maker unknown,
United States
Purchase, 1951

This sleigh is believed to have belonged to the Revolutionary War hero General Peter Gansevoort of Albany, New York.

reported vehicle was the two-wheel pleasure carriage, which comprised 62% of the total number of vehicles reported for Westchester County.

In favorable weather travel by boat on the Hudson River to New York City was probably preferred to coach travel. The road to New York City was usable but crude and the jolting of the coach necessitated frequent repairs.[9] The number of sloops and vessels reported for tax purposes in this county suggests that transportation by water continued to be popular into the nineteenth century.

Albany County: A region of economic growth

The population of Albany County in 1800 was 34,043, slightly more than half the population of New York County. Its principal city, Albany, was the capital of the state from 1797. Although the combined total of real estate and personal property values was high, percentage vehicle ownership was a scant .4% although 21% of the vehicles reported were in the $300-to-$800 category. This apparent disparity is probably attributable to the presence of large landowners, such as the Van Rensselaers, whose land was worked by tenant farmers. Private pleasure

vehicles, particularly the higher-valued vehicles, were more than likely the property of the landowners.

Albany County, centrally located in the eastern area of the state and bordered by the Hudson and Mohawk Rivers, was ideally situated to become an important trade and transportation center. In fact, Albany County provides an excellent example of the role of the horse-drawn vehicle in the development of a transportation system and in the quickening pace of economic activity. Albany County was settled by the Dutch in 1621, taken over by the English in 1664 and established as one of New York State's original counties in 1683. Before the American Revolution the county's primary economic asset was its beaver and fur trade. By the last decade of the eighteenth century the city of Albany was at the center of a metamorphosis that would affect the entire county. The initial impetus for this transformation was the mass migration of New Englanders into the area, which began in the 1780s and continued for three decades. A 1795 report claimed that 500 horse-drawn sleighs carried migrants into the city and that 1,200 horse-drawn sleighs passed through the city during another three-day period.[10] Albany's unpaved and poorly lit streets could not meet the demands of this increased traffic, and agitation for improvement fell on responsive ears. Street paving began in 1793 but took until 1869 to complete. Lighting was improved by the use of more street lamps; their number increased from 20 in 1771 to 586 in 1828.[11] The arrival of additional settlers and the continuous stream of migrants stimulated economic development, as is evidenced by the growth in the number of stores–from 70 in 1781 to 131 in 1796. Increasing numbers of ships carried the produce of the interior from Albany to New York City via the Hudson River. In 1795 there were 90 vessels plying the river between the two cities; wheat was the most important item of cargo.[12] Before the vessels could be loaded, the goods had to be brought to the docks in horse-drawn wagons and sleighs.

Enterprising men in Albany sought to enrich themselves and their community by building networks of roads to expedite the transport of goods and people. In 1785 the New York State legislature granted Isaac Van Wyck, Tallmadge Hall and John Kinney a 10-year exclusive right to operate a stage line on the east side of the Hudson River. In 1790 Ananias Platt obtained a similar right to open a stage line between Albany and

Lansingburgh; this line was extended to Bennington, Vermont, the next year. By 1795 Platt ran his stage six times daily. Of course, the stage was horse-drawn.[13] The next phase in the development of a transportation and commercial network was the building of turnpikes, including the following: in 1797, from Albany to Schenectady; in 1798, from Albany to Lebanon; in 1799 the Great Western; in 1805, from Albany to Delaware County; in 1806, from Troy to Schenectady.[14]

Albany became a center for early stage and mail routes as well as for turnpikes diverging in all directions and extending as far west as Buffalo:

Probably there is no spot in the United States where so many public stages meet and find employment as at Albany. They issue from thence upon every point in the compass. (American Traveller, *1827)*[15]

The horse-drawn vehicle not only facilitated the development of Albany's transportation system, it also provided economic opportunity for stagecoach owners and drivers, tavernkeepers (since taverns sprang up along the turnpikes), turnpike owners and opera-

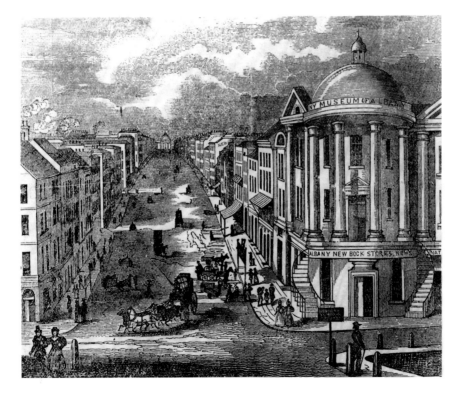

State Street, Albany
1837
Albany, New York
Photograph Courtesy of
McKinney Library,
Albany Institute of
History and Art,
Albany, New York

tors, wheelwrights and carriage makers. James Goold, for example, came to Albany in 1813 as a wheelwright and remained to become a premier manufacturer of stagecoaches, carriages, sleighs and railroad cars. John Butterfield, who was born in 1801 in Bern, Albany County, and whose first job was with a stagecoach line rounding up passengers at boat landings and canal stops, later founded the American Express Company.[16] By 1855 there were 29 coach and wagon manufacturers in Albany County, with 412 employees, a capital investment of $190,560 and a cash value of manufactured articles of $502,823.[17]

Although the War of 1812 interrupted the thrust of internal improvements, it underscored the need for a statewide system of transportation. At the conclusion of the war, plans for the construction of the Erie Canal began. When completed in 1825, the Erie Canal effectively linked the far-flung areas of New York State from Albany to Buffalo on Lake Erie; the Hudson River linked Albany and New York City, so Albany was truly at a commercial crossroads. The success of the Erie Canal stimulated the building of other canals. Outlying regions of New York State used navigable waterways to dig canals that connected with the Erie Canal. Other states followed suit, and New York State then had access by waterway to the Ohio River Valley and to the Mississippi River. An intricate canal system made it possible to travel by boat through New York State to the Great Lakes area and into the Midwest and vice versa. Long-distance hauling by wagon had always been vastly more expensive than shipping freight by water. The developing canal system supplanted wagons for long-distance freight but, as the railroad would soon do, created a new demand for shorter-range transportation by horse-drawn vehicles of both people and goods. Railroad building began in 1831 with the Albany to Schenectady line.

Clinton And Ontario Counties: New York State's Frontier

Clinton County, unlike Albany County, was newly established, having been created in 1788 from Washington County. Bordered by Canada on the north and by Lake Champlain on the east, Clinton was a rugged wilderness. Settlement was promoted by Zephaniah Platt and 32 associates who purchased 31,500 acres and resold them for $1 to $2.50 an acre.[18] In 1798 Plattsburgh was a small settlement without streets and with only four or five roads. In 1800 the county's four towns had a population of 4,359.[19] Internal road building did not begin until 1806 and was interrupted by the War of 1812. The two enumerated vehicles listed on the tax abstracts were probably owned by Platt or one of his associates. To most of the other inhabitants such vehicles were not necessary, whereas livestock, horses and slaves were considered essential.

Ontario County was also a newly-established county, created from Montgomery County in 1789. It was a huge fertile area located in the westernmost part of the state. The first settlements were on Lake Ontario, and according to the federal census of 1790 there were only 1,081 inhabitants in the entire county. The fertile land, however, attracted settlers, and by 1800 the population was 15,218 in 19 towns. Demographically, it was also a young county, with 3,023 free males under 10 years of age and 2,630 free females under 10 years of age.[20] There were no enumerated vehicles in this area, where livestock and horses were investment priorities. Lake Champlain and Lake Ontario served as the most convenient conduits for the transport of goods and people for the inhabitants of Clinton and Ontario, respectively. More than likely, however, horse-drawn sleighs and farm wagons were utilized to reach these waterways. Of course, horse-drawn vehicles also provided some settlers with the means of reaching these developing areas.

Summary, The Early Years

The counties of New York, Richmond, Westchester and Albany were original counties formed in 1683. Whether these counties were densely or sparsely populated by 1800, a portion of their inhabitants could trace their American roots back to the seventeenth or early eighteenth century. Thus by 1800 sufficient time had elapsed for individuals or families to acquire property and position and to establish communities, as well as to accumulate material goods, often including privately owned taxable horse-drawn vehicles. In contrast, both Clinton and Ontario typify settlements in an early stage of development with time, effort and money being invested in erecting houses, clearing the land, establishing farms and building a community. Privately owned horse-drawn pleasure vehicles were super-fluous items in this environment.

New Yorkers were proud of their evolving transportation network. As early as 1809 the Albany Almanac cited the 67 turnpikes in New York State, the 3,071 miles of road, the 21 bridges and the 88 incorporated turnpike road and bridge companies with capital of more than $5.5 million as evidence of the "great and rapid progress in prosperity, in enterprise, in population, in agriculture, in commerce, in wealth, in strength, and in natural resources."[21] This was an optimistic period when change was possible and profitable, when everyday conditions could be made somewhat more amenable and when the road map of New York State was altered by human endeavor, assisted and inspired by the horse-drawn vehicle.

THE MIDDLE YEARS, 1850 TO 1870
Introduction: New York State in the 1860s

By 1860 there were 33 railroad lines in New York State. Steam-powered boats had begun to ply its waters, the telegraph facilitated communications and the gradual introduction of gaslight throughout New York State made travel and living conditions more pleasant. These improvements facilitated the exchange of goods.

Internal improvements were only one aspect of New York State's growth. The population in 1860 was nearly six times greater than in 1800. Natural increase had been augmented by migration from other parts of the nation in addition to immigration from Europe, especially Ireland and Germany. The population spread throughout the state into 60 counties. In addition, a fundamental demographic shift in the population was discernible. In 1800, New

Yorkers were predominantly engaged in agriculture, with commerce an important pursuit in the New York City area. The 1800 distribution of the population reflected this orientation, with 514,000 people living in rural areas and 75,000 in urban areas. By 1860 these figures were 2,356,000 and 1,524,000 respectively, and in 1870 population distribution was almost even, with 2,193,000 inhabitants in rural areas and 2,189,000 in urban areas.[22] Urbanization and industrialization were well underway during these middle years. By 1860 New York State was a leading producer of both agricultural and manufactured goods, as well as the preeminent financial and commercial center of the nation.

The population explosion of New York State from 1800 to 1860 supplied the labor to build the transportation network and to produce the goods for an expanded market. New Yorkers were themselves a vast market for the consumption of goods. The natural resources of the state–navigable waters, minerals and fertile land–were also important elements in the growth of this period. Capital, in the hands of enterprising individuals and assisted by a cooperative legislature, financed internal improvements.

The process of growth was neither even nor consistent. There were periods of financial boom and bust, and some regions developed at a slower pace than did others. Towns like Rochester grew overnight while others were bypassed. As the wheat-growing area moved farther west, farmers relocated, turned to other crops or occupations or became impoverished.

By 1860 the nation was on the verge of a civil war. When hostilities broke out in 1861 the federal government needed men, materials and money. In 1862 the United States Congress enacted widespread tax legislation to meet wartime financial demands. Taxes were levied on income, on the production and distribution of all goods and merchandise, on the ownership of certain items of personal property and for licenses required for all occupations and professions. Once again privately owned horse-drawn vehicles were subject to taxation:

- A $1 tax was placed on one-horse vehicles valued at $75 and over

- A $2 tax on two-horse vehicles valued from $75 to $200
- A $5 tax on two-horse vehicles valued above $200 but not over $600
- A $10 tax on two-horse vehicles valued at more than $600.

Sleighs and vehicles used exclusively for farming or the transport of merchandise were not taxed (Appendix A). Stagecoaches, omnibuses and livery vehicles were taxed, however, as they had not been in 1799 and 1800.

Tax records for the counties of New York, Richmond, Westchester, Albany, Clinton and Ontario may be utilized to ascertain specific information. How many taxable vehicles were there in each county? Who owned the vehicles? What types of vehicle were owned? Who owned more than one vehicle? How many livery stables were there? What was the nature and extent of public horse-drawn transportation? How did the horse-drawn vehicle fare in this period when other modes of transportation such as steamboats and railroads were available?

Unlike the earlier tax records, these records identify the owners of horse-drawn vehicles and the number of vehicles they reported owning. Unfortunately, most of the vehicles are listed merely as one- or two-horse wagons or carriages rather than by the specific type of vehicle.[23] A number of the persons listed in these tax records were researched further for information about their socioeconomic status, and most of this detailed information is contained in Appendix B.

New York County and Environs: Growth of a Metropolis

New York County was the taproot of New York State's growth. By 1860 this county was the state's and the nation's financial center, most important port and most populous city. It encompassed the 22 wards of New York City situated on Manhattan Island, an area 13.5 miles long and 2.5 miles wide at its broadest part, totaling 22,000 acres. In 1800 a population of 60,459 was concentrated on the southern end of the island below Canal Street. By 1860 the population had increased to 813,728, and

The West Side Waterfront, New York City
September 4, 1869
A. R. Waud
From Harper's Weekly
Photograph Courtesy of John Grafton, New York in the Nineteenth Century
Dover Publications

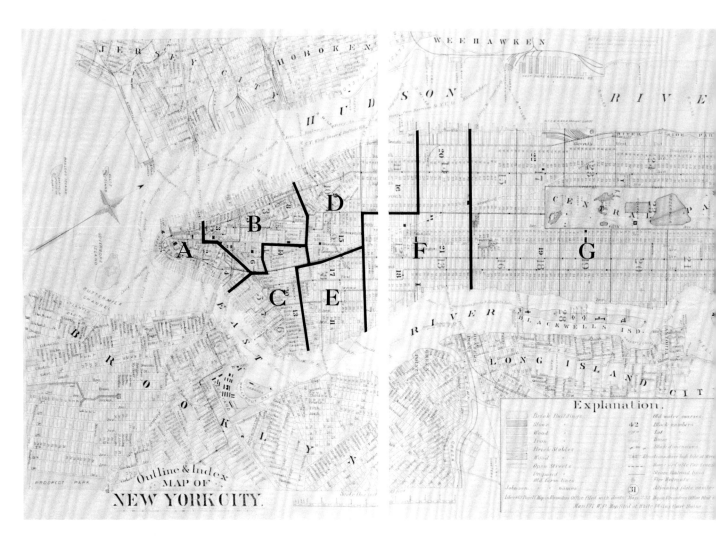

Outline and Index
Map of
New York City
1885
From Atlas of the City of
New York
Editor, E. Robinson
Photograph Courtesy of
The Richard H. Handley
Long Island History Room,
The Smithtown Library,
Smithtown, New York

Key to Map of New York City

A. Financial and
Commercial District
(Wards 1, 2 and 4)

B. City Hall and SoHo
(Wards 3, 5, 6 and 8)

C. Lower East Side
(Wards 7, 10, 13 and 14)

D. Greenwich Village
and Chelsea
(Wards 11 and 17)

E. From 14th Street to
40th Street, excluding
Chelsea
(Wards 18, 20 and 21)

G. Uptown
(Wards 12, 19 and 22 and
Randall's, Blackwell's and
Ward's Islands)

was spread throughout the island. The composition of this population was diverse. In fact, in 1865, naturalized voters outnumbered native-born voters by 77,475 to 51,500.[24]

The inhabitants of the city pursued a wide variety of occupations. There were in 1865, for example, 3,998 butchers, 937 grocers, 6,307 boot- and shoemakers, 1,985 saloon keepers, 3,627 tobacconists, 5,978 merchants, 1,232 lawyers, 1,269 physicians, 1,608 teachers, 6,352 carpenters, 910 boiler makers, 338 gold- and silversmiths, 1,348 brokers, 325 bankers, 33,282 servants, 21,231 laborers, 17,620 clerks, 439 civil officers, 1,546 policemen and 218 farmers. Those engaged in carriage-related occupations included 650 coach and wagon makers, 1,895 drivers and coachmen, 252 livery stable owners, 298 wheelwrights and 2,621 blacksmiths. The salary scale for wage

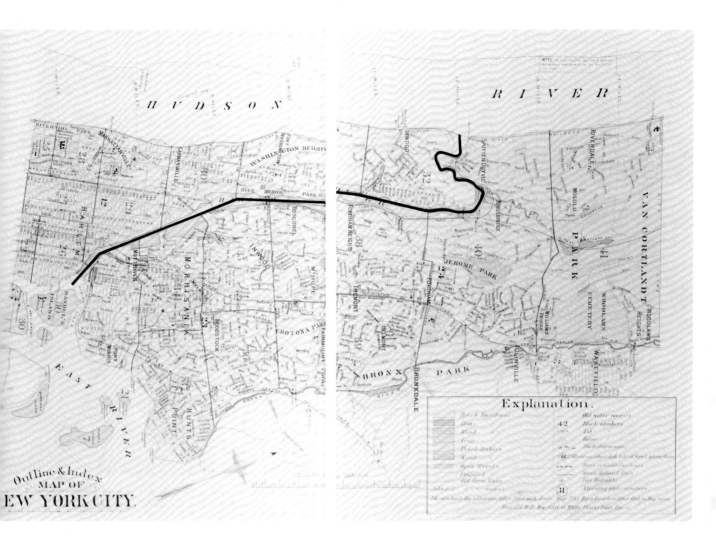

employees was the highest in the state. Men earned from $12 to $128 per month and women earned from $8 to $60.[25]

The city itself was one of contrasts, with areas that were congested, noisy and odoriferous as well as those that were serene and pleasantly rural. There were densely populated areas and sparsely populated areas. The wealthy occupied elegant homes while the poor lived in hovels. There were financial areas, livestock markets, factories, small shops, grand emporiums, ironworks and a multitude of foundries.

The Financial and Commercial District (Wards 1, 2 and 4)[26]

Banks, insurance companies, trading companies, brokerage houses, ferry slips, docks and warehouses were located in this area. In addition, there were three city-owned markets, which were leased to individual retailers and wholesalers. This district also encompassed "newspaper row," where James Gordon Bennett started the *Herald* and Horace Greeley founded the *Tribune*. Another celebrated individual, P. T. Barnum, enlivened the area with his theater at 218 Broadway. In 1850 his star attraction was Jenny Lind, "the Swedish Nightingale." Governor's Island was a fortification with 105 guns, a garrison for 800 men and a training ground for enlistees.[27]

The district served as a terminal point for omnibus lines, stage lines and horse-drawn city railroad lines. From lower Manhattan ferry connections could be made to Brooklyn, Queens County, Staten Island and New Jersey. In this busy commercial district, few residents owned taxable vehicles. Of the 20 vehicles reported for tax purposes, 11 were owned by three livery stables. Two individuals owned another five vehicles and four individuals each owned one vehicle (See Appendix B for details of ownership).[28]

City Hall and SoHo (Wards 3, 5, 6 and 8)[29]

This area was a varied one, encompassing the City Hall section, SoHo, the notorious five points intersection, two vast city-owned markets, the Astor House Hotel and A. T. Stewart's famed "Marble Palace" department store.

SoHo, a modern acronym for South of Houston Street, had undergone several changes in the 20 years before 1860. In 1840, it had been a highly fashionable hotel and shopping area, which by 1850 had given way to an uneasy mix of fine hotels and shops surrounded by dance halls, brothels and casinos. From 1860 to 1890 the area was dominated by factories and warehouses. Nevertheless, the Astor House Hotel, erected by John Jacob Astor in 1835, could boast that it was gaslit and had baths on every floor.

The notorious Five Points intersection was the juncture of Worth, Baxter and Park Streets; contemporaries referred to it as the "den of thieves" or as "Murderers' Alley." It was a slum complete with street gangs, a huge brewery, dilapidated houses and all the other trappings of a neglected neighborhood. Shortly before 1860 renovation of this area began with the demolition of the brewery and most of the other buildings.

The residents of this district were, with some notable exceptions, on the lower rungs on the economic ladder, and the ownership of taxable vehicles reflects this condition. Of the 217 reported vehicles, 124 were owned by 21 livery stables. Twenty-three individuals owned 54 of the remaining vehicles and 39 individuals owned one vehicle each. Two women owned vehicles. One, Sarah Ryder, a widow, owned a livery stable with two carriages taxed at the rate of $2 each. The other, Mrs. Dodge, had one wagon taxed at $1. Vehicle owners in this area included merchants, physicians, fruit and fish retailers, grocers, piano manufacturers, jewelers, brewers and liquor dealers, but such examples do not tell the whole story here: Only 62 individuals in this district owned vehicles for private use, while the vast majority of the 92,196 residents could not afford them. (Appendix B).

Carriage-related businesses in this district included: Schmidlin & Driscoll at 135 Mercer Street, which manufactured coach lamps. Reported value in September 1862, $58.33; in October 1862, $142.60.

Lord & Taylor Dry Goods
c. 1865
Broadway and Grand Street, New York, New York
The Edward C. Arnold Collection
The Metropolitan Museum of Art, New York, New York
Photograph Courtesy of Museum of the City of New York

DeVoursney Bros. at 46 Green, which manufactured coach lamps. Reported value in December 1862, $789.75.

Minor & Stevens at 72 Walker Street, which manufactured carriages. Reported value in September 1862, $2,530; in December 1862, $985; in May 1863, $20,249.

William H. Gray at 27 Wooster, which manufactured carriages. Reported value in December 1862, $1,125; in May 1863, $1,850. Gray advertised in the Commercial Register as follows:

Manufacturer and dealer Coaches, Hearses, every description of Light and Heavy Wagons made to order. Also on hand, an assortment of new and second hand vehicles. Jobbing neatly attended to.

Philip Ketterer at 90 Thompson. Reported value in November 1863, $175. He advertised in the Commercial Register as follows:

Light carriages, Fancy Grocery, Express, and all kinds of Business Wagons made to order on the shortest notice. Particular attention given to manufacturing Express Wagons of all sizes. All work made from the best material, and warranted. Repairing in all its branches, done in the best manner.[30]

These two advertisements attest to the proliferation of types of vehicles available at this time. Further, the customer had the choice of a made-to-order vehicle or of an in-stock new or used vehicle. Vehicles used for occupational purposes were exempt from tax; thus, information about a widely-used form of horse-drawn vehicle is not available from the tax records.

Other carriage manufacturers were D. J. Dusenbury at 102 Laurens and Wood Bros. at 74 Cortland Alley. Charles Swift manufactured wagons at 44 Mercer Street. R. O'Callaghan at 98 Elm made horse collars and reported $100 for each month from May to August and $150 for both September and October. Gibson Wood at 362 Broadway made harnesses and reported $2,651 for December 1862.

The Lower East Side (Wards 7, 10, 13 and 14)[31]

Another mixed area of the successful and the struggling, of the indigent, the industrious and the infamous, was the Lower East Side, which was rapidly becoming the home of a vast number of immigrants. In contrast, William B. Crosby, who had inherited a great deal of land in this district from his father-in-law, Henry Rutgers, still resided in this area and reported an income of $20,595.43, and property of 725 ounces of silver and two carriages, one taxed at $1 and one taxed at $5. Another resident was William M. Tweed, the Tammany Hall figure known as "Boss" Tweed; in the 1863 New York City Directory he is listed as a lawyer and a county supervisor. He reported owning one carriage taxed at $1.

Among the successful businessmen were John Hecker and George Hecker, flour millers and wholesalers of baked goods. John Hecker's income was reported at $43,792.89; he owned one carriage taxed at $1 and one taxed at $2, as well as two coaches taxed at $5 each. George Hecker's income was $50,715.18 and he had two carriages taxed at $1 and one taxed at $5. Alanson Briggs, a manufacturer and retailer of barrels, had an income of $13,585.80 and owned three vehicles, one wagon taxed at $1, one wagon taxed at $2 and one carriage taxed at $5. A more modest and probably more typical businessman for this area was Joseph McKee of McKee & Sons, retailers of groceries and liquor. His income was $569.33 and he owned one carriage taxed at $1. One woman, Mrs. Margaret Willett, owned one carriage taxed at $5. She was a widow of Marinus, a gasfitter whose income was reported at $5,052.46. Other residents who owned vehicles were physicians, lawyers, builders, brewers, carpenters, butchers and candle makers.

Excluding the vehicles used for public transportation, only 66 vehicles were reported for private use. Of these, 30 vehicles were kept by individuals owning one vehicle each. Another 15 individuals reported a total of 36 vehicles. In this district, as in the previous district, few residents in a densely populated area owned vehicles for private use.

Public transportation facilities, however, abounded. There were 26 livery stables with 114 vehicles, including most of the coaches in the district and a barouche. For example, James Gillespie's livery stable maintained three wagons taxed at $1, $2 and

Grand Street Livery Stables
c. 1865
New York, New York
Photograph Courtesy of The New-York Historical Society, New York, New York

$5, one barouche and two coaches each taxed at $2 and three coaches taxed at $5. A stage line with 28 omnibuses, owned by John O'Keefe and John Duryea, operated out of 351 and 353 Rivington Street. In addition, there were five carriage and wagon manufacturers and four harness makers located in this district.

Greenwich Village and Chelsea (Wards 9, 15 and 16)[32]

As indicated in Table 2, the residents of this district owned 1,337 taxable vehicles. In terms of totals, 28.5% of the vehicles were owned by livery stable keepers and stage lines, 24% were owned by individuals who kept just one vehicle for private use and 47.5% were owned by individuals who kept more than one vehicle for private use. A breakdown of the information in Table 2 illuminates the differences in the patterns of ownership in these three wards.

	GREENWICH VILLAGE		CHELSEA
Wards	**9**	**15**	**16**
Total Vehicles	201	750	386
One-horse wagon	1	3	–
One-horse carriage	128	284	179
Two-horse carriage			
at $2	35	193	62
at $5	35	247	141
at $10	2	23	4
Livery stables:			
Number of establishments	13	25	12
Number of vehicles	57	156	86
Stage lines:			
Number of lines	–	–	2
Number of vehicles	–	–	81
Number of individuals owning more than one vehicle	26	159	50
Number of vehicles owned	55	448	131
Individuals owning only one vehicle	89	146	88
Number of women owning one or more vehicles	2	23	2

Fifty-six percent of the vehicles were in Ward 15, 29% were in Ward 16 and 15% were in Ward 9. Slightly more than 59% of the vehicles taxed at $5 and $10 were owned by residents of Ward 15, where, in addition, a greater number of women owned vehicles and a greater number of individuals owned more than one vehicle.

The two wards, 9 and 15, that constituted Greenwich Village exhibit markedly different characteristics, as seen in the above table. The westernmost section of Greenwich Village appears to have been much less affluent than the area around Washington Square. Among the vehicle owners in the West Village were butchers, undertakers, grocers, carpenters and physicians.

Ward 15 encompassed the Village proper, including Washington Square Park and the University of the City of New York, now New York University. This ward was one of the city's most fashionable residential locations, including among its residents such wealthy and influential men as William B. Astor, Cornelius V. S. Roosevelt, "Commodore" Cornelius Vanderbilt, Lloyd Aspinwall, Samuel Lord, A. T. Stewart, Peter Lorillard, James Lenox and William C. Rhinelander. In 1850 a row of private stables was built in MacDougal Alley for the use of the wealthy residents of the area. B. F. Breiden, a merchant with a business location at 105 Chambers Street and a resident of the posh Brevoort Hotel, owned two carriages taxed at $5 and $10 respectively. Albert Clark, the proprietor of the Brevoort, reported one carriage taxed at $1 and two carriages at $10.

Chelsea (Ward 16) began to be developed in the 1820s by Clement Clarke Moore, a scholar and poet best known for "A Visit from St. Nicholas" ("T'was the night before Christmas"). His grandfather Captain Thomas Clarke had purchased a sizable tract of land in 1750 and named it Chelsea. Moore inherited this land and decided to develop it as a residential area free of alleys, stables and manufacturing.[33] Implementing this plan, he donated land for the construction of the General Theological Seminary. As a landowner Moore was apparently able to establish the guidelines for the neighborhood; in the 1830s his friend Don Alonzo Cushman built a row of Greek Revival townhouses opposite the General Theological Seminary.

This pleasant and peaceful environment changed in due course. In 1860 the quiet was disturbed by the Hudson River Railroad, which ran on Eleventh Avenue, and by the Eighth Avenue horse-drawn city railroad. In 1860 the resident owners of vehicles were physicians, lawyers, builders, book dealers, brokers, masons, grocers and brewers (Appendix B).

Tompkins Square Area (Wards 11 and 17)[34]

Dwellings, manufacturing establishments and wholesale and retail businesses commingled in this densely populated area. Residents were merchants, physicians, apothecaries, grocers, bakers and liquor dealers. Those engaged in manufacturing produced cigars, candy, clothing, hoop skirts, furniture, barrels, horseshoes and even steam engines. The manufacture of gun carriages and coffins in the 1860s serves as a grim reminder of the Civil War.

This district encompassed the newly created Tompkins Square Park, several cemeteries and two city markets, one of which was a livestock market. It was also the site of the American Bible Society, established in 1852, McSorley's Ale House, which opened in 1854 as an all-male establishment and remained so until recently, and Cooper Union, founded in 1859 by Peter Cooper as a free coeducational nonsectarian college for the advancement of science and art. Located along the East River were coal- and lumber-yards and the Manhattan Gas Light Company. In 1865 this company produced 400,000 feet of gas each day, lighting 30,000 street lamps and serving 50,000 dwellings and businesses.[35]

The manufacture of horse-drawn vehicles was extensively pursued in this area, with 14 such enterprises listed in the tax records. Many producers of horse-drawn vehicles specialized in the manufacture of a particular type of vehicle. A. Kipp & Sons is identified as a manufacturer of truck wagons, carts and wagons; the term "truck" was already in use to describe a particular type of vehicle.

Three stage lines were based in this

area: Macheral & Simpson, with 27 vehicles (one wagon taxed at $1, one wagon taxed at $2, 14 carriages taxed at $2 and 11 carriages taxed at $5); W. N. Pullis, with 27 vehicles (14 omnibuses taxed at $2 and 13 omnibuses taxed at $5); William Siney with 22 vehicles (one carriage taxed at $1, 10 omnibuses taxed at $2 and 11 omnibuses taxed at $5). The route of each omnibus line was painted on the outside of the vehicle and the fare charged depended on the line used regardless of the distance traveled on that line. Fares ranged from 4 cents to 5 cents.

Of the 315 vehicles reported for tax purposes, 166, or 53%, were owned by livery stables and stage lines. Individuals owning one vehicle account for 48 vehicles, or 15%. Thirty-nine individuals reported owning 101 vehicles, or 32% of the total. In all, 87 individuals—less than 1% of the population of this area—reported vehicles for private use. Most of the vehicle owners resided on Second Avenue between Rivington Street and 14th Street.

Individuals who reported owning vehicles in the most expensive category (taxed at $10 each) were the following: A. V. H. Stuyvesant, 6 carriages: 1 at $1, 4 at $5, 1 at $10. Abraham Deurke, 3 carriages: 1 each at $1, $5 and $10. R. B. McGlovance, 2 carriages: 1 at $2, 1 at $10. L. Comeau, 3 carriages: 1 each at $2, $5 and $10. E. Kettletass, 2 carriages: 1 at $5, 1 at $10. Conklin's Livery Stable, 10 carriages: 5 at $1, 2 at $2, 2 at $5 and 1 at $10.

The individuals owning these expensive vehicles also reported the possession of silver and billiard tables. The only woman listed, Helen Stuyvesant, reported two carriages, one taxed at $2 and one taxed at $5, and 500 ounces of silver.

From 14th Street to 40th Street, excluding Chelsea (Wards 18, 20 and 21)[36]

This area was both the most populous, with 174,058 inhabitants in 1860, and the area for which the greatest number of vehicles, 2,304, was reported. As indicated in Table 2, only carriages were listed, with many of them taxed at $5 and $10. It was also the area with the highest number of women reporting the ownership of vehicles.

In terms of totals, 232 of the vehicles, or 10%, were owned by livery stables and stage lines; 486 individuals owned one vehicle each, or 21%; 574 individuals owned 1,586 vehicles, or 69%. In all, 1,060 individuals–less than 1% of the total population of this district–kept vehicles for private use (Appendix B).

Ward 18 (Gramercy Park area): In 1860 the Gramercy Park section of this ward probably qualified as New York City's most exclusive residential area. Established in 1831 by Samuel B. Ruggles, it was a series of elegant townhouses surrounding a fenced private park owned by the surrounding residents, located at East 20th and East 21st Streets between Third and Fourth Avenues. The Fifth Avenue section of this ward also had its share of the City's elite in residence; August Belmont and Moses Taylor were among the first inhabitants of this section of Fifth Avenue when it opened for residential occupancy in 1851. An 1860 account in the *New York Coach-Maker's Magazine* describes the private stables and coach-houses built on Fifth Avenue and its side streets for the residents of this area. These were not simple structures but were made of brick or freestone with interior gaslight and water from the Croton Water Works. The account refers to the stables and coach-house of "a wealthy banker, lately a foreign minister"–a striking resemblance to August Belmont. In this individual's stable there were "half a dozen splendid coach and saddle-horses" housed in "large airy stalls" while the coach-house contained "three or four rich heavy coaches of European manufacture and an American phaeton."[37]

The ward was also the site of two gasworks. The New York Gas Light Company, with works at 21st and 22nd Streets from First Avenue to the East River, manufactured, washed, purified and distributed gas to the streets below Grand Street. The Manhattan Gas Light Company at 18th Street was capable of supplying 2 million feet of gas in a 24-hour period.[38] In addition, at least 11 horse dealers were located on 24th Street.

Of the total number of vehicles reported for these three wards, 1,108 or 48%, were owned by the residents of the Gramercy Park ward. Nineteen livery stables

Town Coach
c. 1860
Maker unknown,
United States
Purchase, 1955

This elegant closed vehicle
was designed for private use
in the city, and would have
been driven by a coachman.
It was expensive to purchase
and to turn out such a
vehicle.

maintained 83 vehicles, or 7.5%; owners of single vehicles accounted for 225 carriages, or 20.5%; and 285 individuals owned 800 carriages, or 72%. There were 419 carriages taxed at $1,267 at $2,344 at $5 and 78 at $10. Here again ownership reflected the wealth and status of the residents who were bank presidents, bankers, brokers, owners of various types of firms, lawyers, politicians, merchants, importers and physicians. For the most part these individuals maintained offices or businesses in lower Manhattan. Besides Belmont and Taylor, Peter Cooper, Theodore Roosevelt, Samuel Tilden and Hamilton Fish resided in this ward.

Ward 21 (North of 26th Street, East of 6th Avenue, and South of East 49th Street): Belle-vue Hospital, at 25th Street and the East River, was located in this ward, as were cigar factories, coal- and lumberyards and ironworks. Like the Washington Square and Gramercy Park areas, however, this ward included the residences of some of New York's wealthiest individuals, such as John Jacob Astor Jr., William Astor, R. R. Stuyvesant and Charles Tiffany. Far less wealthy individuals–a butcher, a plumber and a builder–also lived here and owned vehicles.

Residents of this ward reported 932 vehicles, or 40.5% of the total for the district. Of the vehicles listed, 362 were taxed at $1,199 at $2,305 at $5 and 66 at $10. Thirteen livery stables maintained 64 carriages, or 7%; owners of single vehicles accounted for 184 carriages, or 19.5%; and 248 individuals owned 684 vehicles, or 73.5%. Ownership of vehicles in this ward reflected the wealth and status of the owners.

Ward 20 (West of 6th Avenue, between 26th and 40th Streets): In addition to residences, this ward contained a hog market at 37th Street and the Hudson River, cattle pens, coal yards and factories for the manufacture of soap and candles.

Of the three wards, the residents of this ward reported the fewest vehicles: 264, or 11.5%. There were 124 carriages taxed at $1, 91 at $2, 47 at $5 and 2 at $10. Of the 264 vehicles, 12 livery stables maintained 41 carriages and one stage line had 44 vehicles, or 32%; owners of single vehicles accounted for 77 carriages, or 29.5%; and 41 individuals owned 101 vehicles, or 38.5%.

Uptown (Wards 12, 19 and 22 and Randall's, Blackwell's and Ward's Islands)[39]

The uptown areas of New York County, north of 40th Street from the

Hudson River to the East River, had large expanses of vacant land. A sizable portion of this land had already been acquired for the development of Central Park. At 42nd Street and Sixth Avenue, Reservoir Park was the receptacle for the city's water supply from Croton Dam in Westchester County. In back of this park, America held its first World's Fair, the Crystal Palace Exhibition, in 1853. Latting Tower, built for the fair, had an ice-cream parlor on the first floor and a viewing area on the top floor that was reached by steam-powered elevators.

This district also contained the city's largest livestock market. Located at 44th Street and Fifth Avenue, this market covered 14 acres with 150 stockyards whose stalls could hold 10,000 animals. Nearby, the cars and horses for the Sixth Avenue City Railroad Line were kept. Farther north were a hog market, a horse exchange, slaughter-houses, an armory and the facilities for the cars and horses of the Third and Eighth Avenue City Railroad Lines.

In 1860 Turtle Bay, on the East River, was noted for its choice residences with their high stoops. Farther up the East River, from 79th to 90th Streets, prosperous middle-class Germans resided in Yorkville. Several other distinct neighborhoods were located in this district. Fort Washington, on the heights overlooking the Hudson River, was the site of impressive dwellings. Manhattanville, on the Hudson River from 125th to 132nd Streets, also contained fine dwellings. In addition it was the site of a Roman Catholic convent, school and college. Iron, paints and refined sugar were produced in this section of the district. Harlem, between Eighth Avenue and the East River at about 106th Street, was an important manufacturing district for the production of indian rubber, chemicals, candles, carriages, ale and beer. The Harlem Gas Light Company, located at First Avenue and 110th and 111th Streets, produced 4,486,834 cubic feet of gas in November of 1862.[40]

Steinway and Sons, located at Fourth Avenue between 52nd and 53rd Streets, manufactured and sold pianos reporting a gross income of $39,376.05 in December 1862. Twelve carriage and wagon manufacturing firms were also located in this district. W. C. J. Dunn reported four coaches valued at $2,800 in September 1862 and three coaches valued at $2,250 in October 1862. For the first time, the tax records reveal a manufacturer's value for a coach, $700 to $750. The selling price was undoubtedly considerably higher.

Of the 191 vehicles reported for tax purposes, 96, or 50%, were owned by livery stables and stage lines; 25, or 13%, were kept by individuals who had only one vehicle; and 21 individuals reported owning a total of 70 vehicles, or 37%. Thus, only 46 individuals owned taxable vehicles–an imperceptible percentage of the population of these wards, which was 125,171 in 1860. Not unexpectedly, the most conspicuous vehicle owners resided in Fort Washington and Manhattanville; James Gordon Bennett was among those who resided in Fort Washington (Appendix B).

Few of the reported vehicles were identified by type. Those that were include the buggy, the top wagon, a light wagon, a spring wagon, the rockaway, the barouche, the phaeton, a chaise, and a coupé. The one woman in this district who reported a vehicle, Susan King, owned a phaeton taxed at $2.

There were several transportation facilities in this area. Marshall and Perry operated a stage line with 30 stages, based at West 45th Street. A steamboat line, the Harlem and New York Navigation Company, reported December 1862 receipts of $2,452.81. There were also five horse-drawn railroad lines, as follows:
• Second Avenue Railroad, which ran from 42nd Street and Second Avenue down to Peck Slip, returning to 23rd Street and Second Avenue and then up to Harlem. Receipts for September 1862 were $21,705.31.
• Third Avenue Railroad, which ran from Ann Street in lower Manhattan to the Harlem Bridge and back. Fare was 5 cents to East 65th Street and 6 cents to 130th Street. Receipts for September 1862 were $52,115.70.
• Sixth Avenue Railroad, which ran from Vesey Street to West 59th Street and back. Fare was 5 cents. Receipts for September 1862 were $28,982.70.
• Eighth Avenue Railroad, which ran from Vesey Street in lower Manhattan to

West 70th Street and back. Fare was 5 cents. Receipts for September 1862 were $31,022.23.

• Ninth Avenue Railroad, which ran from Barclay Street in lower Manhattan to West 54th Street and back. Fare was 5 cents. Receipts for September 1862 were $8,539.86.

These north-south routes connected the residential and commercial areas of the county and provided a means of traveling to and from the ferries to neighboring Richmond, Long Island and New Jersey. The receipts collected in just one month, as indicated above, confirm their popularity and frequent usage. The Ninth Avenue Line, with the lowest receipts, was the most recent addition to New York County's public transportation system. The Third Avenue Line, with the highest receipts, listed capital of $1,170,000 in the New York City Directory of 1863. By 1865 this line utilized 90 horse-drawn cars. All lines were owned, operated and maintained by private companies.

Of the 4,592 vehicles in New York County in 1862, 25%, or 1,151, were owned by livery stables or stage lines. Nine hundred fifty-five individuals reported owning one vehicle or 20.5%; another 909 individuals reported owning a total of 2,486 vehicles, or 54.5%. Only 1,864 individuals–less than one-quarter of 1% of the county's population–owned vehicles for private use.[41]

Within New York County, the concentration of taxable vehicles was in those areas where the wealthy resided, and the owners usually reported more than one vehicle as well as silver plate and billiard tables. The accumulation of their wealth derived from inheritance, business activities, capital investments and real estate. They tended to live in the developing areas of the city where their neighbors were much like themselves. For the most part they were descendants of families whose American ancestry could be traced back to the seventeenth or eighteenth century. (A. T. Stewart and August Belmont, who were nineteenth-century arrivals, had a head start on other immigrants in the form of capital and connections.) While middle-class residents could afford and did own vehicles, the nature of their ownership pales in comparison with the number and value of the vehicles owned by the upper class.

Women who owned vehicles were the wives or widows of prominent men or descendants of early New York families. Whole segments of the population, servants and laborers, for example, did not own vehicles kept exclusively for private use. In New York County the ownership of such vehicles was neither widespread nor democratized.

Public horse-drawn transportation was extensively developed in New York County, with livery stables, stage lines, omnibuses and horse-drawn railroad lines providing a variety of ways of traveling from one area to another. Probably the fares were generally beyond the means of the working class but not of the middle class.

Richmond County (Staten Island)

This county's population had increased by 1860 to slightly more than five and one-half times its population in 1800. The principal industries were the oyster trade, the manufacture of dyes and prints and the production of beer. Hourly ferries to New York City enabled Richmond's inhabitants to work in that city if they so chose. In turn, New Yorkers sometimes established residences in Richmond.

All vehicles are listed in the tax records for Richmond as either one-horse or two-horse carriages and are not identified by specific type. Livery stable keepers owned 10% of the vehicles, individuals with only one vehicle accounted for 29% and individuals owning more than one vehicle accounted for 61%. In all, 302 individuals–slightly less than 1.25% of the population–maintained 531 vehicles for private use.

Five vehicles, owned by three individuals, were reported at the $10 tax figure. Lucius Tuckerman, an iron merchant (premises in New York City) whose income was $35,000, owned a carriage taxed at $10 as well as two carriages taxed at $1. Mrs A. C. Meyer, with a reported income of $1,000, owned one carriage taxed at $10, and one taxed at $5, as well as 1,016 ounces of silver. Agatha B. Mayer owned three carriages taxed at $10, six taxed at $2 and one taxed at $5, as well as 655 ounces of silver.

In addition to the two women mentioned above, 16 other women reported owning carriages. One of them, Mrs. William Burger, is listed as a livery stable keeper with four vehicles. Seven other women owned a total of 20 vehicles. Eight women reported owning one vehicle each. Six of the women–one-third of the total number of women with vehicles–reported incomes ranging from $300 to $5,400. The number of women compared to the number of men owning taxable vehicles is small: only 5.5% of the private vehicle owners were women, who reported 7.5% of the private vehicles in Richmond.

Other vehicle owners include Dwight Townsend, a member of the United States Congress from 1863 to 1865, who had three carriages, one taxed at $1, one taxed at $2 and one taxed at $5. John Steers, a manufacturer and retailer of furniture and coffins, owned one carriage, taxed at $5. Jacob H. Vanderbilt owned five carriages; three taxed at $1, one taxed at $2 and one taxed at $5. Vanderbilt also declared 119 ounces of silver, a billiard table and an income of $11,606.40. William Henry Vanderbilt owned four carriages, two taxed at $1 and two taxed at $2; although 160 ounces of silver are also listed, there is no listing of this Vanderbilt's income.[42] A. F. Ockenhausen, with a reported income of $70,641, kept four carriages, two taxed at $1, one taxed at $2 and one taxed at $5.

Although Mr. Ockenhausen's income is the highest, many other incomes were reported at five figures. The salary range for wage employees was relatively high; men earned from $20 to $90 per month, with the highest number of wage earners, 141, listed at the $50-per-month rate. Women's wages ranged from $9 to $26 per month. An impression of affluence is conveyed, and it is reinforced by the number of servants in this county (1,679), the number of drivers and coachmen (142)[43] and the frequent declaration of the possession of silver plate. The names of many vehicle owners indicate that they were descendants of those recorded in the 1709 Census, with some variations in the spelling of the names.[44]

Carriage-related businesses include 14 coach and wagon makers, nine livery stables, 21 wheelwrights and the drivers and coachmen previously mentioned. There do not appear to have been railroads, horse-drawn

or otherwise, on Staten Island in 1860; an 1835 attempt to establish such service had been abandoned. In 1883 a horse-drawn railroad was operating on the north shore; by this time this area was also served by the Richmond County Gas Light Company, which provided gaslight for private customers. Only Edgewater had gaslit street lamps, as did one area of New Brighton.[45]

Westchester County

In 1860 Westchester County was a thriving area with a population three and one-half times greater than its 1800 population. The county's inhabitants were settled in 24 towns and scattered in numerous small villages. Steamboats on the Hudson and East Rivers carried the county's farm and manufactured goods to nearby New York City and Connecticut as well as to more distant markets. The construction of three railroad lines that traversed Westchester stimulated the establishment of towns and villages along these lines. In those areas closest to New York City, skilled workingmen could commute by steamboat or railway to jobs in that city.[46]

Many of Westchester's residents were engaged in growing vegetables and fruit, in dairy farming and in the manufacture of firebricks and iron products. In addition there were 1,306 carpenters, 160 railroad men, 5,385 laborers, 820 clerks, 5,165 servants, 225 dressmakers, 907 merchants, 145 brokers, 17 publishers, 197 lawyers, 122 physicians, 303 teachers, 174 clergymen and 146 tobacconists. Carriage-related businesses fared well in this county; there were 109 coach and wagon makers, 490 drivers and coachmen, 84 wheelwrights and 15 livery stables. Wages for men ranged from $12 to $70 per month, with 650 listed at $50 per month. Women's wages ranged from $6 to $26 per month.[47]

As indicated in Table 3, Westchester was second only to New York County in the number of taxable vehicles reported. In terms of totals, 815 individuals each owned one vehicle, accounting for 45% of the vehicles; 359 individuals had more than one vehicle, for a total of 919, or 51%; the remaining 4% were owned by livery stables and stage lines. Thus, only 1,164 individuals–slightly less than 1.25% of the population of this county–maintained vehicles for private use (Appendix B).

The figures for Westchester in Table 3 include seven nonresidents who had country homes in that county. These individuals reported a total of 15 vehicles, which were maintained at their Westchester residences. Two examples will suffice. John D. Wolfe of New York County listed two wagons, one taxed at $1 and one taxed at $2. Lorillard Spencer of New York County reported two carriages taxed at $1, one taxed at $2 and one taxed at $5, as well as a $10,000 yacht and a billiard table.

Sixty-six women are listed as vehicle owners; only in New York County did more women declare vehicles. Like their male counterparts, the women of Westchester owned a variety of vehicles, with a preference for the rockaway. Catherine Spencer owned a one-horse rockaway, a two-horse rockaway taxed at $5, a two-horse rockaway coupé taxed at $5 and a two-horse barouche taxed at $5. She also reported an income of $29,431.72. Most of the women who owned vehicles were listed as married, with no indication as to whether or not they were widows.

Collectively and individually the inhabitants of Westchester owned a wide variety of vehicles, including the ubiquitous buggy, spring wagons, top wagons, phaetons, barouches, landaus and several types of rockaways. Coaches were used as both private and public means of transportation.

Public forms of horse-drawn transportation were widely available in this county, and livery stable owners kept on hand a variety of vehicles suitable to the tastes and needs of their clientele. Henry Foster, for example, had five buggies taxed at $1, two wagons taxed at $1, one barouche taxed at $1, two rockaways taxed at $2 and one omnibus taxed at $2. Although the tax records do not contain any specific reference to stage lines as such, it seems that at least two individuals provided this service. Edward Baack, of Morrisania, owned three omnibuses, each of which was taxed at $2, and Henry Baack, of Belmont, owned six stagecoaches, each of which was taxed at $2. More than likely they used these vehicles for some form of public transportation.

Four hotels evidently maintained horse-drawn vehicles for the convenience of their guests. A North Salem hotel owned by Henry L. Crosby had a stagecoach taxed at $2. Another in West Farms owned by Jefferson Lewis utilized three omnibuses taxed at $2 each. Two hotels in Mount Vernon used hackney coaches. One, owned by Richard Woods, kept two hackney coaches taxed at $2 each. The other, owned by George Gould, had one hackney coach. It was not uncommon in this period for the better hotels to provide transportation for their guests to and from the railroad station or the steamboat dock.

Summary: New York, Richmond and Westchester Counties

The size of the population in the three counties just discussed provided a ready market for their products, enhanced by the ease of transportation from one area to another. In addition, individuals could and did reside in one county and work in another; the beginnings of a metropolitan area are discernible.

A comparison of the 1860s statistics with those of 1800 reveals a dramatic decrease in per capita ownership of horse-drawn vehicles in New York County and an even more dramatic increase in per capita ownership in Westchester County. The rapid growth in population, the widening gulf between rich and poor and the availability of public transportation explain the change in New York County's pattern of ownership. The affluence of Westchester accounts for the increase in ownership there. The per capita ownership of Richmond County decreased slightly. All three counties are well represented in the carriage categories, especially with regard to those in the $5 tax bracket. The most commonly reported vehicle in each county, however, was the one-horse carriage; it is important to remember that all one-horse vehicles valued at $75 and over were taxed at $1. An elegant one-horse vehicle with custom trimmings, therefore, was subject to the same rate of tax as the plainest one-horse vehicle, unless the latter was valued at less than $75, in which case it was not taxed and does not even appear in the tax records.

Albany County: The Crossroads of New York State

In 1860 Albany County encompassed an area of 544 square miles, smaller than its 1800 expanse, since Schenectady had been established as a separate county. The population of Albany County had more than tripled, to 113,917. In addition, the population was more varied, as the arrival of Irish and German immigrants had diversified the previously predominantly Protestant, Dutch and English population.

The city of Albany was the county's most important city. As the capital of New York State and the county seat, it contained government buildings, an almshouse, a hospital, a penitentiary and an observatory. A scrutiny of the Albany City Directory for 1860 confirms that the area realized the potential for economic growth discussed earlier. By 1860 there were 232 retail grocers, 76 attorneys, 100 physicians, 30 hotels, 50 boardinghouses, 20 livery stables, 13 newspapers, seven libraries, eight express companies, 13 public schools, 17 banks, three insurance companies, nine sewing machine companies (including the Singer Sewing Machine Company), two gasworks companies, two telegraph companies, three stage lines, nine railroad lines, four turnpike companies, three steamship lines and daily omnibuses, to name but a few of the many businesses and professions represented. There were also 15 carriage makers, including James Goold, and five coach lace weavers. Five of the carriage makers advertised on-premises "carriage smiths."[48]

Despite the increased population and economic growth, the number of taxable vehicles reported for the entire county is surprisingly small–1,578. This figure includes 142 vehicles used for public transportation, or 9% of the total number of vehicles. Sixteen livery stables kept 137 vehicles, two individuals each owned two omnibuses and one hotel had an omnibus. Some 159 individuals owned 371 vehicles, or 23.5%. Another 1,065 individuals each owned one vehicle, or 67.5%. Thus, 1,224 individuals–slightly more than 1% of the population–owned vehicles for private use.

Vehicles that are specifically identified include 476 buggies, seven rockaways, six omnibuses, eight coaches, seven barouches, one stanhope, one caleche, three chariotees and three hearses. The residents of Albany were decidedly partial to the one-horse buggy and other one-horse vehicle types. Four of the listed vehicles were taxed at $10; two of these were hearses owned by Augustus Brewster's livery stable in the city of Albany. Another was a coach owned by Worthington Foland's livery stable in Albany. The fourth was a carriage owned by Joel Rathbone, also of Albany. Six of the eight coaches reported were owned by livery stables in Albany. Harriet S. Bennett, a retail liquor dealer, owned a coach taxed at $2, and John F. Rathbone, a manufacturer and wholesale dealer of stoves, owned a coach taxed at $5. Mr. Rathbone, unlike Harriet S. Bennett, also had two buggies, one taxed at $1 and one taxed at $2.

Several of Albany's vehicle owners, including Joel Rathbone, Erastus Corning and Thomas W. Olcott, were connected with most of the important enterprises and institutions in the city of Albany. They were in a position to influence the course of events in the city and beyond. The tendency in this period of businessmen to diversify their economic pursuits was pronounced in Albany. The concentration of influence in the hands of the few was not novel, but the type of individual wielding power was now the businessman or the financier rather than the landowner (Appendix B).

Few women other than Harriet S. Bennett owned taxable vehicles, even though the Directory lists women as dressmakers, millinery retailers, nurses and teachers. Their economic status was a major reason that few women owned vehicles; in Albany, wages for women ranged from $6 to $20 per month. Even wealthy women, however, like Mrs. Stephen Van Rensselaer, were not recorded as owning vehicles. Mrs. Van Rensselaer was engaged in benevolent activities, serving as the second directoress of the Albany Guardian Society and Home of the Friendless. She certainly could have afforded a carriage, so economics was not the only operative force in vehicle ownership. Local social attitudes may have been equally important in limiting the number of women owning vehicles in their own right. The response to the women's suffrage movement

by the women of Albany may shed some light on the prevalent attitudes. From 1854 to 1861 a series of women's suffrage conventions was held in Albany. Elizabeth Cady Stanton, Susan B. Anthony and Ernestine Rose attended and addressed these conventions, apparently drawing little support from local women. According to Howell and Tenney's *History of Albany,* "all the eloquent addresses of these women never aroused even a quartet of women in Albany to organize for woman's suffrage work."[49] None of the three local supporters of this movement, Mrs. Thompson, Lydia Mott and Phebe Jones, is listed as a vehicle owner.

Augustus Brewster's livery stable contained nine one-horse carriages, one two-horse carriage, one chariotee, two coaches, two barouches taxed at $2, two barouches taxed at $5, one caleche and two hearses. Obviously the customer could choose both the type of vehicle to hire and the quality of the vehicle, as is evident from the listing of the barouches at different tax rates. While not all the other livery stables offered as extensive a variety as did Brewster's, they did have some selection of vehicles on hand.

In addition to hiring vehicles, residents and visitors could use either omnibuses or stages. In 1860, omnibuses ran daily from Albany to the Lumber District, to Kenwood, to Mt. Hope, to Newtonville and to Troy.

Stages ran from Albany to Berne, New Scotland, Bethlehem, Waterlo, Rensselaerville, Clarksville, Nassau, DeFreistville, Sandlake, New Baltimore, New Lebanon, Schoharie and Charlotteville. In 1863 the first horse-drawn street railway service began in the city of Albany, operating from South Ferry via Broadway to the Lumber District. It is possible that this line replaced the omnibus service to the Lumber District. In 1864, a second horse-drawn street railway was put into service on State Street.[50]

Horse-drawn transportation in Albany coexisted at this time with accessible travel by steamboat and railroad. This in itself is an impressive change from 1800, when horse-drawn vehicles provided the only alternative to walking, riding horseback or using a boat. Although private vehicle per capita ownership increased between 1800 and 1862, ownership in 1862 was not widespread; perhaps the ready availability of horse-drawn vehicles for hire and the omnibuses and stage lines lessened the need for private ownership of taxable vehicles.

Clinton And Ontario Counties: The Rural Periphery

Clinton County by 1860 was considerably reduced in territory, since the counties of Franklin and St. Lawrence had been formed from its 1800 expanse. Clinton County, which consisted of 1,092 square miles bordered by Canada on the north and Lake Champlain on the east, occupied the northeastern corner of New York State. Its rugged terrain disguised rich and extensive iron-ore deposits, which had proven to be an invaluable asset to the county, and numerous streams and falls were an excellent source of water power for lumber mills. Clinton's wilderness areas still abounded with animals whose fur provided another prime resource. Lake Champlain served as the link to the Erie Canal. The Champlain Canal, completed in 1823, facilitated the transport of people and goods from Clinton to Albany and from there to points in western New York and beyond, as well as down the Hudson River to New York City. The combination of natural resources and the means with which to transport them stimulated the development of this county.

By 1860 the population of Clinton County was 45,735, nearly 10 times greater than it had been in 1800. There were 14 towns. Plattsburgh, with a population of 6,630, was the county seat, with civic buildings, the Plattsburgh Gas Light Company, the Iron Bank, the Mercantile Bank, the Northern New York Mutual Insurance Company, several mills, a foundry and two tanneries. Champlain, with a population of 5,859, had a linen factory, a carriage factory and two tanneries. Dannemora, whose population was 1,371, was the site of the Clinton State Prison, where convict labor was used to mine and manufacture iron. There were two iron mines in this town, ten in Au Sable, and two in Black Rock.[51]

Occupations included the raising of livestock for meat, leather goods and dairy products. In 1864 2,236 neat cattle were killed for beef and 6,204 swine were slaughtered for pork. Fourteen tanneries were in operation. Dairy farmers made 946,725 pounds of butter and sold 100,020 pounds of cheese. Poultry farmers realized $10,812.17 from the sale of eggs. Collectively, other farmers harvested 605,368 bushels of potatoes, 438,016 bushels of oats and 96,072 bushels of apples. Other products produced in this area were maple sugar, starch, linen, spring wheat, Indian corn and honey. There were 4,482 farmers, 1,957 laborers, 240 miners, 165 sawyers, 35 lumbermen, 206 carpenters, 639 servants, 226 clerks and accountants, 46 dressmakers, 11 bankers, 16 lawyers, 49 physicians, 226 clergymen, 114 teachers, 170 merchants, 26 grocers and 36 peddlers. The pay scale for men working for wages ranged from $24 to $68 per month, and for women from $16 to $20.[52]

Individuals employed in carriage-related occupations included nine carriage manufacturers, 63 wheelwrights and four drivers and coachmen. H. L. Isham, listed as a carriage manufacturer in Plattsburgh, reported the following values:

September	1862	Carriages	$155.
September	1862	Wagons and Sleighs	$ 93.
February	1863	Wagons and Sleighs	$568.
April	1863	Wagons and Sleighs	$495.
May	1863	Carriages	$705
June	1863	Carriages	$440.

Common-Sense Buggie
1862
From catalog, Lawrence, Bradley & Pardee
New Haven, Connecticut

These figures reflect the seasonal impact on demand, as is verified by the May 1863 notation that "wagons and sleighs now sold or moved for consumption." The harsh winters in this area resulted in a long selling period for sleighs and wagons and a short one for carriages.

In view of the foregoing, it is not surprising that there were so few taxable horse-drawn vehicles reported. The number of one-horse vehicles in each category listed in Table 3 exceeds that of two-horse vehicles. One-horse vehicles account for slightly more than 71% of the total number of vehicles–289 of the 405; the one-horse buggy was the most popular vehicle. Three hundred twenty-four individuals reported the ownership of just one vehicle–80% of the total, and only 31 individuals reported owning more than one vehicle, accounting for 66 vehicles–16.5% of the total. Only 355 individuals altogether–three-quarters of 1% of the population–kept taxable vehicles for private use.

Individuals with more than one vehicle included Zephaniah C. Platt, who had both a one-horse and a two-horse carriage. Mr. is listed in the tax records as an insurance agent with an income of $1,184.14. Other biographical information indicates that he was the first president of both the First National Bank of Plattsburgh and of the Clinton County Savings Bank, as well as a promoter of the move to build the New York and Canada Railroad. Mr. Platt was the grandson of Charles Platt, who had moved to Plattsburgh in the early years of its settlement and who was the brother of Zephaniah Platt, one of the founders of Clinton County.[53] Roswell O. Barber, a wholesale dealer with an income of $3,048, owned two one-horse carriages. Laura Nye, a widow, owned a one-horse buggy, a two-horse wagon taxed at $2 and a two-horse carriage taxed at $5. Edmond Kingsland of Au Sable owned two wagons taxed at $1 and one wagon taxed at $2. He manufactured horse nails and iron castings. Peter Keese, also of Au Sable, had two buggies taxed at $1 and one carriage taxed at $2. He declared an income of $6,151. Lorenzo D. Larkins (or Laskins) of Beekmantown, a retail butcher, reported one wagon taxed at $1 and one taxed at $2.

In addition to Laura Nye, six other individuals owned vehicles taxed at $5: George Sevinance (a retailer of lumber), George Hoyle, Timothy Hoyle, A. S. Thirbin, Theodorus Platt and Mrs. Richard Hayworth. Sevinance reported an income of $2,125.50. George Hoyle's income was $2,985.10, while Timothy Hoyle's was $2,097.93. Platt reported an income of $98.18. The incomes reported by the inhabitants of this county are generally low in comparison with those of the other counties covered in this essay; the low level of income is probably the reason why there were so few vehicles reported at the $5 tax rate and none at the $10 level.

The one omnibus listed was owned by George Angill, a hotel and livery stable keeper in Champlain. This vehicle probably was used to transport his guests to and from the railroad or steamboat. The coach owned by Benjamin C. Webster, the proprietor of another hotel in Champlain, may have served the same purpose. A two-horse stage was owned by Jefferson Bishop of Dannemora and another was owned by James Johnson of Au Sable. These vehicles were probably used for local public transportation.

Clinton County's two railroads, the Plattsburgh-Montreal and the Champlain-St. Lawrence, reported receipts of $719.82 and $113.51, respectively, in September 1862.

Long severe winters probably explain why the inhabitants of Clinton County owned so few taxable vehicles as well as why their vehicles were apparently used primarily for local purposes. Long-distance travel was more than likely made by boat or railroad. On reaching one's destination one could hire a vehicle from a livery stable or use some form of public transportation such as a local stage or an omnibus.

Ontario County in 1800 had encompassed most of the western part of New York State, but by 1860 it was much smaller since the counties of Genesee, Livingston, Monroe, Wayne and Yates had been formed out of its original territory. The reorganized Ontario County covered an area of 640 square miles, with a population slightly less than three times greater than that of 1800 settled in 15 towns.

Although smaller in size than Clinton County, Ontario County was more fertile. In 1864, its 5,727 farmers harvested 659,820 bushels of winter wheat, 410,301 bushels of oats, 874,394 bushels of Indian corn, 359,126 bushels of potatoes, 694,512 bushels of apples and 190,854 bushels of barley. In addition, 14,377 swine were slaughtered for pork and 3,253 neat cattle killed for beef. Dairy farmers made 1,110,592 pounds of

butter and sold 119,357 pounds of cheese. The income from the sale of eggs was $27,218.86. [54]

Besides farmers, workers in other occupations included 1,937 laborers, 163 merchants, 311 carpenters, 148 clerks, 101 physicians, 168 teachers, 1,161 servants, 179 boot and shoemakers and dealers, 69 lawyers, 87 clergymen, 12 bankers and 12 peddlers. Engaged in carriage-related industries were 96 coach and wagon makers, five drivers and coachmen, 11 livery stables and two wheelwrights. Men who worked for wages earned from $18 to $65 per month, while women earned from $8 to $15. [55]

Canandaigua, situated on Lake Canandaigua, was the county seat, with the usual civic buildings, churches, schools, the Bank of Canandaigua, the Canandaigua Gaslight Company and three newspapers. A daily steamer arrived from Naples, at the opposite end of the lake. The town was also a station on the New York Central Railroad line. Population was 7,075 in 1860. Geneva, situated at the foot of Seneca Lake, had a population of 5,057 in 1860 and could boast of nine churches, public and private schools, the Hobart Free College, three newspapers, the Geneva Gaslight Company, daily steamer service and a connection to the Erie Canal via the Cayuga and Seneca Canal. [56]

Ontario County is, in microcosm, an excellent source for studying the reform movements that swept the nation in general in the antebellum period. It was deservedly called the "burned-over district;" its citizens were involved in religious revivals, temperance, women's rights, abolitionism and the workingmen's movement. [57] It was also the scene of the alleged kidnapping and probable murder of William Morgan by a group of Masons, which precipitated the Anti-Masonic movement and the Anti-Masonic political party. Both men and women were active in most of these movements. Elizabeth Blackwell, the first American woman to receive a medical degree, graduated from Ontario's medical school. [58]

The inhabitants of Ontario County owned more than three times as many taxable vehicles as did the inhabitants of Clinton County, whose population was approximately the same. A possible expanation incorporates a combination of factors, such as a more favorable climate, a higher standard of living, a less rugged terrain and a more dynamic society. Also noteworthy is the fact that 52 women owned vehicles in this county. Only New York County and Westchester County listed more women owning vehicles; perhaps the activist nature of this county created an environment conducive to such ownership.

Another unique characteristic of the pattern of ownership in this county is the number of one-horse carriages reported. Eighty-one percent of the total number of vehicles listed in the tax records were one-horse carriages. One-horse carriages were owner driven and usually more affordable than two-horse vehicles; perhaps Ontario's populace purchased this type of vehicle because they valued these characteristics. Such an explanation would seem consistent with the extremely high percentage of owners of single vehicles. As indicated in Table 3, 1,154 individuals reported owning only one vehicle–84% of the vehicles listed. In comparison, a mere 86 individuals reported owning more than one vehicle, accounting for 179 vehicles, or 13%. A total of 1,240 individuals, or slightly more than

2.75% of the population, reported owning vehicles for private use (Appendix B).

Few carriages are identified by type except for the coach owned by a Mrs. Denton, taxed at $5, and a coach owned by a William Johnson as agent for Denton, taxed at $5. This does not seem to be the same vehicle, since the tax was assessed twice. One other carriage is specifically identified, a one-horse rockaway owned by S. Brush. The single two-horse wagon listed was a spring wagon owned by Albert Chellborg. Finally, 13 buggies taxed at $1 were reported.

In 1800 the inhabitants of Ontario County had no taxable vehicles to report. In 1860, Ontario County ranked fourth in terms of the number of taxable vehicles reported by the inhabitants of the six counties discussed in this essay and on a per capita basis, Ontario County ranked first.

Summary of the Middle Years

Of the 10,340 taxable vehicles reported in the six counties covered by this essay, 44.4%, or 4,592, were in New York County.

Six-seated Germantown Rockaway
January, 1860
From The New York Coach-maker's Magazine, *Edited and published by Ezra M. Stratton New York, New York*

The buggy, the most frequently identified vehicle, was obviously popular. Since other types of vehicles are so infrequently specified in the tax records, it is difficult to draw conclusions about the popularity or the use of any particular type of vehicle. An article in the *New York Coach-Maker's Magazine*, however, sheds some light on the types of vehicles favored by New Yorkers.[59] Among the vehicles mentioned in this account are the coach, the caleche, the barouche, the phaeton and the rockaway. The coach is referred to as "the richest and most respectable of private carriages," costing from $1,000 to $1,500, and the author notes with some derision the growing tendency to display coats of arms and crests on the coach panels. The popularity of the caleche is attributed to its "removeable" top, which rendered it a vehicle for all seasons. This type of vehicle cost from $900 to $1,000. The barouche, a large open carriage costing from $600 to $800, was favored by those who could afford more than one carriage and by women because the carriage could accommodate their hoop-skirted dresses. Both the

phaeton and the rockaway are described as less expensive, smaller, simpler vehicles. The phaeton appealed to those who enjoyed driving their own carriage while the rockaway suited those who "desired the convenience of a carriage without ostentation." Confirming the impression conveyed by the tax records for this county, the account asserts that New York was "getting rich rapidly," as was evident from "the surprising number of elegant equipages seen in the streets." Carriages were not merely a mode of transportation but were also a reflection of the wealth, status and taste of their owners.

Livery stables and stage lines accounted for 14% of the total number of vehicles in the six counties, or 1,475. Another 4,484 vehicles, or 43.5%, were owned by those who had just one vehicle. Some 1,675 individuals owned a total of 4,381 vehicles, or 42.5%; thus, a total of 6,159 indivduals, or less than 1% of the population of the six counties, owned vehicles for private use.

The noticeable drop in per capita vehicle ownership in New York County

Last Horsecar in New York City
July 26, 1917
Photograph Courtesy of Museum of the City of New York, New York, New York

This horsecar was on i farewell run, on Bleec Street, New York City.

between 1800 and the 1860s is attributable to several factors. Primary among these was the county's rapidly increasing population, most of whom were at the low end of the economic scale, too poor even to ride the public transportation, much less to own a vehicle and maintain it and the horse to pull it. Another factor was the space, or rather the lack of it, in which to keep both horses and vehicles. The wealthy, as previously indicated, built private facilites to accommodate their vehicles and horses. Others had to find room on their own premsies, which was not always feasible, or rent space in a livery stable, thereby increasing the cost of maintenance. Finally, the great improvements in public transportation provided those who could afford to use it with a viable alternative to private vehicle ownership. The middle class probably were the principal users of public transportation. The tax records also indicate that per capita vehicle ownership increased in developing areas like Clinton and Ontario Counties, where distances were greater and where there was no comparable public transportation.

THE FINAL YEARS, 1870 to 1920

Thus far, this essay has dealt with the role of the horse-drawn vehicle in the private and public sectors of society, with the core of evidence derived from tax records. In this section the emphasis shifts to the public sector. Several considerations influenced this shift. First, horse-drawn vehicles were not subject to taxation after the 1870s, so that a comparable quantitative source is not available. Furthermore, the change from horse-drawn to horseless transportation began in the public sector. Finally, the literature of the period predictably concentrates on the newer forms of transportation while ignoring the old. The remainder of this essay is thus limited to an overview of New York and Ontario Counties, which were chosen to illustrate discernible differences in the impetus for change in the modes of transportation that occurred in the last three decades of the nineteenth century.

New York County's population suffered a short-lived decline during the Civil War years, but increased steadily in each subsequent decade: in 1870 to 952,000, in 1880 to 1,209,561 and in 1892 to 1,801,739. The increase and spread of the population outstripped the ability of existing public transportation to meet the county's needs. After the Civil War, steam-powered elevated trains were proposed as a solution to this problem. The owners of horse-drawn modes of transportation and property owners were bitterly opposed, carrying the battle to the courts. Although they admitted the shortcomings of the existing systems–and, indeed, could hardly ignore the loud complaints on all sides about overcrowding, reckless drivers and diseased and overworked horses–and vested economic interests aside, they feared that the elevated lines would adversely affect the environment by darkening the streets, spewing steam and smoke and emitting rattling noise. They also argued that such lines were dangerous to the people and horses below. Despite these objections, the proponents of the elevated lines won the day, and starting in 1870 a series of elevated lines was constructed above the existing horse-drawn railroad routes. Thus the first phase of public horseless transportation supplemented but did not replace horse-drawn transportation.

It was the ability to electrify a transportation system that marked the beginning of the end of public horse-drawn transportation in New York County. In 1882 Thomas Alva Edison opened a generating plant in the county. With this new source of power, the electric trolley and the electric train became feasible. Such transportation was perceived by the public as faster and more comfortable than horse-drawn transportation. The electric cars were larger than those built to be pulled by horses, and the interiors were lit by electricity. From 1887 on, electric trolleys began to replace the horsecars. Nevertheless, horse-drawn public transportation was not abandoned overnight. In 1890, 5,280 horses pulled 2,300 street cars along 268 miles of track.[60] In 1894 horsecar lines, omnibuses and electric trolleys clogged the streets of the county, and various modes of transportation were still overcrowded. By 1915 the elevated railroad and subway system carried 2,468,970 passengers in a single day;[61] public horse-drawn transportation passed from the New York City scene in 1917.

Canandaigua Pier
c. 1910
*Canandaigua, Ontario
County, New York
Photograph Courtesy of
Ontario County
Historical Society,
Canandaigua, New York*

*This photograph depicts fou[r]
distinct modes of land
transportation: horse-draw[n]
carriage, electric trolley,
railroad and automobile.*

In contrast, the natural beauty of the Canandaigua Lake area spurred the development of steamboat transportation for commercial and recreational purposes. In 1867 John and Alexander McKechnie, successful brewers, went into the steamship business. In 1876 they ran a daily service from Canandaigua to Woodville with the schedule geared to coincide with Canandaigua's railroad service. At Woodville, connections to local points were made by stagecoaches.[62] Although this interconnecting system was an effective means of moving people and goods, the horse-drawn vehicle played a subordinate role in the system. The promotional thrust encouraged travel for pleasure to see the natural beauties of the area, reach picnic spots, attend hotel dances, fish, swim and go boating. Travel itself was a form of recreation both on the boat and at its destination. In time the Canandaigua Lake area became a popular resort area, and horse-drawn vehicles continued to be used for short-distance travel and for land excursions.

On September 8, 1887, Canandaigua put in operation a horse-drawn railroad line that ran on Main Street to the railroad station and the docks of Lake Canandaigua. Regular service was from 7:30 a.m. to 9:30 p.m. on weekdays but was adjusted to meet local needs, such as church services on Sundays or performances at the opera house. Initially, two streetcars were pulled by eight horses used in rotation. By 1888, cars ran every 15 minutes in the summer and every 30 minutes in the winter and carried 113,558 passengers. But technology had already advanced beyond horse-drawn cars on tracks. In 1892, Canandaigua suspended operation of its horse-drawn railroad to prepare the way for the installation of electric cars. This costly changeover necessitated ripping up the old tracks and laying new ones, purchasing new cars and building a power plant. Service on the new line began in 1894. The electric trolley made the round trip in 20 to 25 minutes and held as many passengers as several horsecars.[63] The allure of alternative modes of transportation where public horse-drawn transportation had not established itself was irresistible. In such areas the latest technological advance was adopted rather quickly, and at some point the horse-drawn vehicle became identified in the public mind as outmoded and expendable. By 1900 there were 817 street railroads in the United States with 21,907 miles of track for electrically operated lines and only 259 miles of track for horse-drawn lines.[64]

Meanwhile in New York County pleasure driving of privately owned horse-drawn vehicles had become a popular pastime. Central Park, completed in 1864, immediately became the scene of afternoon parades of elegant equipages owned by New York's elite. They came to be seen, to display their carriages and to drive free from the crush of the city's traffic. The general public came to the park in their horse-drawn road wagons and carriages or by means of the horse-drawn railroad,[65] to enjoy the scenic beauty of the park, see the zoo, hear concerts, dance on the mall and court. In the winter, horse-drawn sleighs raced through the snow-covered park.

Another form of pleasure driving began in 1875 with the formation of the New York Coaching Club by James Gordon Bennett and Leonard W. Jerome, among others. This club engaged in competitive four-in-hand driving in vehicles modelled on the earlier English mail coaches. Elegantly attired men and women rode in the coaches' top seats.

The club's annual spring and fall parades imitated earlier stagecoach driving with precisely timed schedules along designated routes. This gentlemen's sport was also a spectator event for those who lined the competition route to watch and to urge the "whips" on. Members of the club included the August Belmonts Senior and Junior, Pierre Lorillard, J. R. Roosevelt, William H. Vanderbilt and Lawrence Jerome. In 1876 coaching with timetables, scheduled routes and paying passengers was inaugurated by wealthy sporting coachmen. For the owners, public coaching was as expensive as private coaching. When the New York Coaching Club considered putting a public coach in operation, the estimated cost was $15,000 for the vehicle, the matched horses and the training of the horses.[66] In 1901 the Ladies' Four-in-Hand Driving Club, in which women did all the driving, was organized. Emulating their male counterparts, the women adopted a club uniform, held annual parades and put a public coach on the road. The members of this club were for the most part the daughters and wives of Coaching Club members. Eleanor Jay Iselin, for example, was the daughter of the Coaching Club's president, Colonel William Jay, and

Jerome Park Race Track
June 19, 1886
Engraving taken from a photograph by Bidwell
From Harper's Weekly
Photograph Courtesy of John Grafton, New York in the Nineteenth Century
Dover Publications

the wife of the Club's vice-president, Arthur Iselin.[67]

The popularity of public and private coaching waned in the 1890s, as documented by the minutes of the Coaching Club. In 1894 the minutes refer to absent members and coaches at the annual events, as well as to fewer spectators. In 1895 the annual spring parade attracted "numerous bicyclists, masculine and feminine, who wheeled up and down the line." In 1907 the spectators arrived "in broughams, victorias, automobiles, phaetons, pony carts." In 1910 the Coaching Club and the Ladies' Four-in-Hand Driving Club held their last spring parades. The reason given for the demise of this event was the "rapid and insistent popularity of the automobile which not only

tended to crowd horse-drawn vehicles off the road, but brought about a complete change in the physical condition of the roads whereby locomotion with horses was uncomfortable and even dangerous." Although the Coaching Club continued to meet, by 1925 membership was "an ever-narrowing circle."[68]

Pleasure driving was not the exclusive prerogative of the clubs or of the wealthy. Individuals of more modest means also enjoyed leisurely driving or riding in their privately owned or rented horse-drawn vehicles. Middle-class vehicle owners drove about in buggies or runabouts.[69] How many individuals owned such private vehicles is difficult to ascertain without a quantitative source that specifically deals with that subject, although national statistics indicate

**Park Drag in Front
of Gerry Mansion
on Fifth Avenue**
c. 1900
New York, New York
*Gift of Minor Wine
Thomas, 1982*

that carriage production peaked during the 1899-to-1909 period. The following chart contains information from the 1910 Census pertaining to the numbers and types of horse-drawn vehicles produced in those years.[70].

	CARRIAGES	WAGONS	PUBLIC CONVEYANCES
1899	904,639	570,428	2,218
1904	937,409	643,755	2,711
1909	828,411	587,685	2,243

The carriage category included family and pleasure vehicles, the wagon category included vehicles made for farm, business and civic use and public conveyances included cabs, hacks, hansoms, hotel coaches and omnibuses. In addition to the figures indicated above, in 1909 there were 14,908 carriages, 42,112 wagons and 104 public vehicles produced by firms engaged primarily in the manufacture of products other than carriages. Conversely, carriage and wagon manufacturers produced 174 automobiles in 1899, 199 automobiles in 1904 and 544 automobiles in 1909. Obviously this was a period of transition and experimentation for those in the business of producing road vehicles. Population statistics indicate that the population of the United States was 75,994,575 in 1900 and 91,972,226 in 1910. Thus in about 1900, slightly more than one pleasure vehicle ("carriage") was produced per 100 people in the population; while in 1910, slightly less than one vehicle was produced per 100 people in the population. It appears that per capita ownership probably did not significantly increase in the final years of the horse-drawn era—at least not on a national basis. Of course, not all of the vehicles produced were for domestic use, for vehicles were also being exported. Nevertheless, in these final years the wealthy tended to own several vehicles with well-defined purposes, just as they had throughout the nineteenth century. The general public was probably buying multipurpose vehicles such as the buggy, or the rockaway, which were being produced in large number and advertised at modest prices in the Sears Roebuck and Montgomery Ward catalogs. In 1902, for example, Sears Roebuck offered open buggies for $27.75 and $28.95; a buggy with a genuine leather quarter top sold for $34.95. By 1908 the impact of the automobile is readily discernible in the Sears Roebuck advertisement of an "automobile" seat top buggy described as "A HIGH CLASS STYLISH UP TO DATE AUTOMOBILE SEAT TOP BUGGY" with a "special design automobile bent front wing dash."[71] This vehicle sold for $59.95. The rapid growth of the automobile industry adversely affected the carriage industry.

After 1909, production of horse-drawn vehicles declined drastically. In 1919, only 216,000 carriages, buggies and sulkies and 324,000 farm wagons, trucks and business vehicles were produced.[72]

THE END OF AN ERA

In 1800 the horse-drawn vehicle had set the pace for transportation and economic development in New York State. By the 1860s the pattern of ownership in terms of the percentage of the population had undergone little change. As regions developed economically, per capita ownership of vehicles increased. Certainly the relationship between individual wealth and carriage ownership is clear. What had changed was the proliferation of the types of vehicles, of carriage makers and of livery stables. Outside of urban areas such as New York County and the city of Albany, public horse-drawn transportation served primarily as the link to railroad and steamship lines or to connect rural areas without railroad or steamship service. Both public and private horse-drawn vehicles were used for local purposes–in particular for the short-distance hauling of

Polo Grounds
c. 1900
New York, New York
Photograph Courtesy of
The New-York Historical
Society, New York,
New York

This was the home of the New York Giants baseball team from 1891 to 1957. There is a noticeable difference between the vehicle types and styles of dress in this photograph and those the illustration of Jerome Park.

Plattsburgh Steam Laundry
*1910-1915
Plattsburgh, Clinton
County, New York
Photograph Courtesy of
Clinton County
Historical Museum,*

of goods to wholesalers and retailers. During the last two decades of the nineteenth century horseless transportation began to replace horse-drawn transportation in the public sector; by 1900 New York State had 96 street railroads with 2,596 miles of track for electrically operated cars and 115 miles of track for horse-drawn cars.[73] In the last three decades of the nineteenth century, especially in New York County, the privately owned horse-drawn carriage was popular as a pleasure vehicle. But another mode of private transportation became prevalent, and by 1890 bicycling was a craze. The bicycle is credited with introducing some of the essentials of the early automobile such as tubular steel, wire wheels, pneumatic tires and the chain-and-sprocket drive,[74] and bicycle riders successfully compelled the county to begin using asphalt pavement. In fact, the preponderance of bicycles brought about a redefinition of the term "carriage" in the *Highway Manual of the State of New York*, as follows:

The term "carriage" as used in this article, shall be construed to include stage-coaches, wagons, carts, sleighs, sleds and every other carriage or vehicle used for the transportation of persons and goods, or either of them, and bicycles, tricycles, and all other vehicles propelled by manumotive or pedomotive power.[75]

In due course, the gasoline-fueled automobile replaced the horse and carriage as the basic mode of private transportation. When, in 1934, the Coaching Club acted favorably on a proposal to write its history, the horse-drawn vehicle was already anachronism. One era had ended, and another begun.

TABLE 1:
Summary of 1800 Tax Records

ITEMS TAXED	VALUE	NUMBER NEW YORK CITY	KINGS*	QUEENS*	SUFFOLK*	RICHMOND	WEST-CHESTER	ALBANY	CLINTON	ONTARIO
Horse-drawn Vehicles:										
Coach	$800	31	–	1	–	–	4	3	–	–
Chariot	$700	61	2	2	–	–	3	2	–	–
Post Chaise	$700	9	–	–	–	–	3	–	–	–
Phaeton on steel springs	$300 }	117	23	14	1	3	16	23	1	–
Coachee on steel springs	$300 }									
Other 4-wheel pleasure carr.	$100	22	32	33	1	15	31	13	–	–
Two-wheel top carriage	$100	136	11	39	15	13	38	16	–	–
Other 2-wheel pleasure carr.	$ 50	289	183	340	241	87	157	77	1	–
Subtotal, No. of Vehicles		665	251	429	258	118	252	134	2	–
Subtotal, Value of Vehicles		$139,150	$21,750	$30,600	$13,950	$8,050	$26,950	$17,450	$350	–
Other Items:										
Ox 4 years & up	$ 15	18	69	1,362 +	3,230	216 +	3,415	1,667	572	3,120 +
Bull 4 years & up	$ 15	2	9	25	28	4	17	23	–	7
Cow 4 years & up	$ 10	617	1,997	6,176	4,802	1,379	7,894	6,251	509	1,938
Neat cattle 3 years	$ 6	9	581	1,918	3,512	447	3,036	2,870	251	1,372
Neat cattle 2 years	$ 4	34	452	1,798	3,051	620	3,256	3,294	322	1,930
Horse or Mare 1 year old	$ 8	10	118	385 +	309	82	673	979	68	187
Horse or Mare 2 years	$ 15	121	143	457	363	81	798	1,089	69	227
Horse or Mare 3 years	$ 20	28	129	453	427	84	697	989	69	104
Horse or Mare 4–8 years	$ 30	625	567	1,725	1,386	336	2,461	3,568	287	779
Gelding/Mare 8 + –12 years	$ 20	663	382	1,041	859	206 +	1,323	1,755	114	511
Gelding/Mare 12 + –16 years	$ 8	219	231	513 +	443	107	749	790	77	236
Stallion or Stud 4 + years	$300	2	12	46	29 +	5	39	67	14	42
Mule 1 year old	$ 8	–	–	–	–	–	1	1	–	–
Mule 2 years	$ 16	–	–	–	–	–	–	–	–	–
Mule 3 years	$ 25	–	2	2	9	–	2	7	–	–
Swine, more than 8, 1 + yr.	$ 3	3	13	181	178	–	508	134	8	210
Brass or Steel wheel clock	$ 40	663	140	190	74	46	109	152	6	10
Gold watch	$ 50	968	49	69	18	11	76	131	3	12
Other watches	$ 12	2,277	304	609 +	590	201	651	765	54	227
Able-bodied slave 12–50 yrs.	$100	929	686	522	346	331	512	862	18	22

		NEW YORK CITY	KINGS*	QUEENS*	SUFFOLK*	RICHMOND	WEST-CHESTER	ALBANY	CLINTON	ONTARIO
ITEMS TAXED	VALUE	NUMBER								
Sloops & vessels 30 + − 60 tons	$500	11	–	5	43 +	–	18 +	7	–	–
Sloops & vessels 60 + tons	$700	–	–	–	–	–	9 +	23 +	–	–
. . . .										
Total Value of Taxed Items										
Enumerated Property		$ 383,280	$ 164,338	$ 315,020	$ 300,424	$ 88,329	$ 421,731	$ 476,932	$ 38,341	$ 145,660
Real Estate		$19,014,203	$1,911,449	$3,620,660	$3,201,401	$590,694	$5,219,991	$5,264,924	$1,153,437	$7,171,336
Residue Personal Property		$ 2,237,252	–	$ 891,696	$ 237,542	$ 39,491	$ 400,752	$ 649,845	$ 3,455	$ 23,883
Total Value		$21,634,735	–	$4,827,376	$3,739,367	$718,514	$6,042,474	$6,391,701	$1,195,233	$7,340,879
Population		60,459	5,740	16,893	19,464	4,563	27,428	34,043	4,359	15,218
Percentage Vehicle Ownership		1.1%	4.4%	2.5%	1.3%	2.6%	.9%	.4%	.05%	.00%

*Not included in subsequent tables and not discussed in text.

TABLE 2
New York County Ownership of Vehicles 1860s Tax Lists

	FINANCIAL & COMMERCIAL DISTRICT	CITY HALL & SOHO DISTRICT	LOWER EAST SIDE DISTRICT	GREENWICH VILLAGE AND CHELSEA DISTRICT	TOMPKINS SQUARE DISTRICT	14TH STREET TO 40TH STREET DISTRICT	UPTOWN DISTRICT	TOTALS NEW YORK COUNTY
Population 1860	42,648	92,196	129,983	117,148	132,524	174,058	125,171	813,728
Population 1865	28,398	71,424	118,269	106,048	138,516	148,166	115,565	726,386
. . . .								
1-Horse Buggy at $1	–	5	–	–	2	–	7	14
2-Horse Buggy at $2	–	1	–	–	–	–	1	2
1-Horse Wagon at $1	–	13	17	4	5	–	23	62
2-Horse Wagon at $2	–	2	9	–	1	–	–	12
1-Horse Carriage at $1	6	92	48	591	94	905	39	1,775
2-Horse Carriage at $2	11	75	88	290	127	557	94	1,242
2-Horse Carriage at $5	1	23	46	423	80	696	17	1,286
2-Horse Carriage at $10	2	6	0	29	6	146	3	192
. . . .								
Totals	20	217	208	1,337	315	2,304	191*	4,592*
Livery Stables								
Number	3	21	26	50	19	44	22	185
Vehicles owned by	11	124	114	299	90	188	66	892

	FINANCIAL & COMMERCIAL DISTRICT	CITY HALL & SOHO DISTRICT	LOWER EAST SIDE DISTRICT	GREENWICH VILLAGE AND CHELSEA DISTRICT	TOMPKINS SQUARE DISTRICT	14TH STREET TO 40TH STREET DISTRICT	UPTOWN DISTRICT	TOTALS NEW YORK COUNTY
Stage Lines								
Number	–	–	1	2	3	1	1	8
Vehicles owned by	–	–	28	81	76	44	30	259
Individuals owning more than one vehicle	2	23	15	235	39	574	21	909
Number of such vehicles	5	54	36	634	101	1,586	70	2,486
Single vehicle owners	4	39	30	323	48	486	25	955
Women owners	1	2	1	27	1	42	1	75
Percentage Vehicle Ownership**	.05/.07	.24/.30	.16/.18	1.14/1.26	.24/.23	1.32/1.56	.15/.17	.56/.63
Percentage Vehicle Ownership not including livery stables and stage lines	.02/.03	.10/.13	.05/.06	.82/.90	.11/.11	1.20/1.40	.08/.08	.42/.47

*Seven vehicles were listed as a group without any differentiation, the tax rate as a total for the seven vehicles was $12.
**Percentage figures were derived by comparing the number of taxed vehicles in 1862 records to population figures for 1860 and 1865.

TABLE 3:
New York State Counties Ownership of Vehicles, 1860s Tax Lists

	NEW YORK COUNTY	RICHMOND COUNTY	WEST- CHESTER COUNTY	ALBANY COUNTY	CLINTON COUNTY	ONTARIO COUNTY	GRAND TOTALS
Population 1860	813,728	25,492	99,497	113,917	45,735	44,563	1,142,932
Population 1865	726,386	28,209	101,197	115,504	45,713	43,316	1,060,325
1-Horse Buggy at $1	14	–	68	476	149	13	720
2-Horse Buggy at $2	2	–	–	3	3	–	8
1-Horse Wagon at $1	62	–	185	79	33	–	359
2-Horse Wagon at $2	12	–	58	347	31	1	449
1-Horse Carriage at $1	1,775	366	771	380	107	1,112	4,511
2-Horse Carriage at $2	1,242	161	576	196	75	232	2,482
2-Horse Carriage at $5	1,286	57	141	93	7	16	1,600
2-Horse Carriage at $10	192	5	3	4	–	–	204
Totals	4,592*	589	1,802	1,578	405	1,374	10,340*
Livery Stables							
Number of	185	11	35	19	9	10	269
Vehicles owned by	892	58	58	142	15	41	1,206

	NEW YORK COUNTY	RICHMOND COUNTY	WEST-CHESTER COUNTY	ALBANY COUNTY	CLINTON COUNTY	ONTARIO COUNTY	GRAND TOTALS
Population 1860	813,728	25,492	99,497	113,917	45,735	44,563	1,142,932
Population 1865	726,386	28,209	101,197	115,504	45,713	43,316	1,060,325
Stage Lines							
Number	8	–	2	–	–	–	10
Vehicles owned by	259	–	10	–	–	–	269
Individuals owning more than one vehicle	909	131	359	159	31	86	1,675
Number of such vehicles	2,486	360	919	371	66	179	4,381
Single vehicle owners	955	171	815	1,065	324	1,154	4,484
Women owners	75	18	66	13	7	52	231
Percentage Vehicle Ownership	.56/.63	2.31/2.09	1.81/1.78	1.39/1.37	.89/.89	3.08/3.17	.90/.98
Percentage Vehicle Ownership not including livery stables and stage lines	.42/.47	2.08/1.88	1.74/1.71	1.26/1.24	.85/.85	2.99/3.08	.78/.84

*Seven vehicles were listed as a group without any differentiation. The tax rate as a total for the seven vehicles was $12.

TABLE 4:
New York State Population Statistics for 1865

	TOTAL NEW YORK STATE	NEW YORK COUNTY	RICHMOND COUNTY	WEST-CHESTER COUNTY	ALBANY COUNTY	CLINTON COUNTY	ONTARIO COUNTY	TOTALS FOR TABULATED COUNTIES
Population 1865	3,831,751	726,386	28,209	101,197	115,504	45,713	43,316	1,060,325
Foreign born	920,293	313,477	9,142	26,394	32,546	12,364	6,657	400,580
Aliens	399,463	151,838	3,956	12,395	10,422	6,094	2,715	187,420
Number of families	790,931	148,683	5,260	19,227	23,288	8,170	8,880	213,508
Owners of land	373,998	11,375	2,441	9,170	7,817	4,813	6,096	41,712
Number of dwellings	594,045	49,844	4,527	15,839	16,427	7,693	8,268	102,598
Value of dwellings	$977,121,378	$423,096,918	$9,129,135	$36,755,440	$27,863,999	$2,696,222	$7,593,653	$507,135,367
Number of unvalued dwellings	66,114	975	374	1,741	1,417	879	753	6,139

NOTES

1 *Historical Statistics of the United States, Colonial Times to 1970* (Washington, D. C., 1975), Part 1, 32 and *Twelfth Census of the United States 1900* (Washington D. C.), vol. 8, 332.

2 The Married Woman's Act of 1848 applied only to property that a married woman inherited or received as a gift. Since a married woman could not sign contracts, she still could not sell, transfer, or bestow her property to anyone in a will without her husband's consent. Nor did she have control of any money she earned. In 1860 the New York State legislature broadened the provisions of the 1848 Act, granting married women full control over any property they received, including any money they earned, and allowing married women to enter into contracts without a husband's counter-signature.

3 The Abstracts of Valuations Made for State Taxes in 1799 and 1800 are located in the New York State Archives at Albany, New York.

4 See "An Act for the assessment and collection of taxes" in *Laws of the State of New York* (Albany, 1887), vol. 4, 402-414.

5 A total of 385 enumerated vehicles is listed in the abstracts for the counties not included in Table 1. The figures are Cayuga 0, Chenango 1, Columbia 82, Delaware 0, Essex 0, Greene 25, Herkimer 6, Montgomery 11, Onondaga 0, Orange 89, Rensselaer 87, Rockland 24, Schoharie 5, Tioga 2, Ulster 32 and Washington 21.

6 Asa Greene, *A Glance at New York* (1837) as quoted by Clarence P. Hornung in *The Way It Was: New York 1850-1890* (New York: Schocken Books, 1977) 33.

7 Edwin Williams, *The New-York Annual Register, 1840* (New York: David Felt, 1840) 389.

8 Richard M. Bayles, ed., *History of Richmond County, New York, From Its Discovery to the Present Time* (New York: L. E. Preston & Co., 1887) 678-683.

9 Harold Donaldsen Eberlein, *The Manor and Historic Homes of the Hudson-Valley* (Philadelphia and London: J. B. Lippincott Company, 1924) 119-120.

10 George R. Howell and Jonathan Tenney, *History of the County of Albany, New York, 1609-1886* (New York: W. W. Munsell & Co., 1886) 311, 610.

11 Howell and Tenney 515-516.

12 Howell and Tenney 609-610.

13 Howell and Tenney 310.

14 Howell and Tenney 311.

15 *The American Traveller,* 12 October, 1827, quoted in Oliver W. Holmes and Peter T. Rohrback, *Stagecoach East* (Washington, D. C.: Smithsonian Institution Press, 1983) 80.

16 David M. Ellis, "The Spirit of Turnpiking," NAHO 12.2 (Summer, 1979):14-18.

17 *New York State Census 1865* (Albany, 1865) 361-365.

18 H. K. Averill Jr., *Geography and History of Clinton County, New York* (Plattsburgh, N. Y.: Telegram Printing House, 1885) 23; Duane Hamilton Hurd, *History of Clinton and Franklin Counties, New York* (Philadelphia: J. W. Lewis & Co., 1880) 152.

19 Hurd 152.

20 *Ontario County, New York State. 1800 Federal Population Census Schedule Transcript and Index* (Cambridge, 1963) vi.

21 *Albany Almanac for 1809.*

22 *Historical Statistics of the United States, Colonial Times to 1970,* Part 1, 32.

23 The manuscript records for the Civil War years are on Microfilm Series M 603 in the National Archives, Washington, D. C. There are 180 rolls of microfilm for New York State. Annual lists are available for 1862 for each of the six counties, with the exception of Wards 1, 2, 4, 18, 20 and 21 in New York County. For Wards 1, 2 and 4, the 1863 list was used, and for Wards 18, 20 and 21, the annual list for 1864 was used. The specific microfilm rolls used in this essay are 114, for Albany County; 91, for Westchester County; 38, for Richmond County; 130, for Clinton County; 171, for Ontario County; 213, for New York County, Wards 1, 2 and 4; 52 and 54 for New York County, Wards 3, 5, 6 and 8; 59 for New York County, Wards 7, 10, 13 and 14; 65 for New York County, Wards 9, 15 and 16; 72 for New York County, Wards 11 and 17; 76 for New York County, Wards 18, 20 and 21; 84 for New York County, Wards 12, 19 and 22. Tables 2 and 3 contain statistical information drawn from these rolls. Table 4 contains population statistics drawn from the 1865 New York State Census.

The scope of taxation and the size of the population made the recording and tabulation of the data a formidable task. The records themselves presented many problems; the handwriting is often illegible, initials are frequently used rather than given names, the data is inconsistently organized and the extant record is incomplete.

24 *New York State Census 1865* 3.

25 *New York State Census 1865* 195-209, 512-518.

26 For tax assessment purposes, New York County was divided into seven districts. Each district encompassed several wards. To ascertain the total number of taxable vehicles reported in this county, it was necessary to record the information contained on seven rolls of microfilm. The organization of this section of the essay follows that of the tax data. In this way, distinctions in the pattern of ownership become clear. See pages 76–77 for a map delineating the wards and their boundary lines.

27 Descriptions of New York County are drawn from several sources, including J. H. French, *Gazetteer of the State of New York* (Syracuse, N. Y.: R. Pearsall Smith, 1860); W. Parker Chase, *New York The Wonder City-1932* with a new introduction by Paul Goldberger (New York: New York Bound, 1983); Gerald R. Wolfe, *New York, A Guide to the Metropolis,* rev. ed. (New York: McGraw-Hill, 1983); Carol von Pressentin Wright, *Blue Guide New York* (New York: W. W. Norton, 1983); D. T. Valentine, *Manual of the Corporation of the City of New York* (New York: Edmund Jones & Co., 1865); and Henry Collins Brown, ed. *Valentine's Manual of Old New York* (New York: Year Book of the Museum of the City of New York, 1925).

28 Throughout this section and in Appendix B, biographical information on the owners of vehicles is drawn from *The Dictionary of American Biography*, Dumas Malone, ed. (New York: Charles Scribner's Sons, 1936) and *Who Was Who in America 1607-1896* (Chicago: The A. N. Marquis Company, 1963).

29 Microfilm Series M 603, Rolls 52 and 54, National Archives, Washington D. C.

30 *Trow's New York Directory*, compiled by H. Wilson (New York: 1863 and 1864), Commercial Register section 11.

31 Microfilm Series M 603, Roll 59.

32 Microfilm Series M 603, Roll 65.

33 Wright 208.

34 Microfilm Series M 603, Roll 72.

35 Valentine 410.

36 Microfilm Series M 603, Roll 76.

37 Ezra M. Stratton, ed., "The Equipages of New York," *The New York Coach-Maker's Magazine*, vol. 2 (May, 1860):47. Those who could not afford private facilities could park their equipages at the local livery stable for a fee, of course.

38 Valentine 410.

39 Microfilm Series M 603, Roll 84. There were no vehicle owners listed for Randall's, Ward's and Blackwell's Islands. These three islands were the sites of institutions such as penitentiaries, almshouses, workhouses, asylums, hospitals, potter's fields and the Emigrant Refuge and Hospital.

40 Valentine 410.

41 Ezra Stratton, *World on Wheels* (New York: by the author, 1878) 458. Stratton estimated the number of vehicles in New York County at 13,562 in 1863. This figure included 5,000 private carriages and wagons, 558 omnibuses, 954 hackney coaches and coupés, 255 express wagons, 416 wood and charcoal wagons, 278 junk carts, 5,374 public carts, and 724 dirt carts. In addition, there were handcarts and sleighs. Note, first of all, that the figure given for private vehicles is quite close to the figure used in this essay from the tax records. Stratton's information evokes a vivid picture of the assortment of vehicle types scrambling for space in the streets of New York County.

42 Background information on some of Richmond's residents was graciously furnished by Mr. Stephen Barto, archivist of The Staten Island Historical Society. Jacob Vanderbilt was the brother of "Commodore" Cornelius Vanderbilt, and William Henry Vanderbilt was the son of the "Commodore." At this time William was a successful farmer on Staten Island, but would soon become his father's right-hand man.

43 *New York State Census 1865* 195-209, 512-518.

44 The 1709 Census is contained in *Staten Island and Its People: A History 1609 to 1929,* by Charles W. Leng and William T. Davis, (New York: Lewis Historical Publishing Co., 1930), vol. 1, 141. Names that appear in the census and in the tax lists include Bodine, Britton, Canon, Clark, Cole, Corsen, Crocheron, Decker, DuPuy, Egbert, Guyon, Jonson, Martling, Mersereau, Merrill, Peryne, Scott, Seton, Siguine, Taylor, Turet, Van Name, Ward, Winan and Journeay. Some of these same names appear in the tax lists as those of peddlers. Decker, DuPuy, Guyon, Egbert, Van Name and Winant are among those listed as taxed for peddlers licenses. Under the tax laws peddlers' licenses cost $20 for those traveling with more than two horses, $15 for those traveling with two horses, $10 for those traveling with one horse and $5 for those traveling on foot. Peddling was a recognized method of conducting business. Here and elsewhere in 1862, peddlers were often residents of the community.

45 French 563-567 and Richard Edwards, ed., *A Descriptive Review of the Manufacturing and Mercantile Industries of the City of Brooklyn, the Towns of Long Island and Staten Island* (New York and Philadelphia: Historical Publishing Co., 1883) 447-448.

46 J. H. French, *Gazetteer of the State of New York* (Syracuse, New York: R. Pearsall Smith, 1860) 696.

47 *New York State Census 1865*, 195-209, 512-518.

48 *The Albany Directory for the Year 1860* (Albany, N. Y.: Adam, Sampson & Co., 1860).

49 Howell and Tenney 735.

50 Howell and Tenney 519. Both lines were still in operation in 1884. By then the South Ferry line had 27 cars, 150 horses and 75 employees, while the State Street line had 44 cars and 215 horses. Running these horse-drawn lines in an urban area required a huge supply of horses that had to be sheltered, fed and attended to.

51 French 232-240.

52 *New York State Census 1865* 195-209, 394-404, 512-518.

53 Hurd 174-175.

54 *New York State Census 1865* 394-404.

55 *New York State Census 1865* 512-518.

56 French 491-500.

57 This term comes from Whitney R. Cross's study of the religious revivals in western New York State, in his work entitled *The Burned-over District: The Social and Intellectual History of Enthusiastic Religion in Western New York–1800-1850* (Ithaca, N. Y.: Cornell University Press, 1950).

58 G. David Brumberg, *The Making of an Upstate Community: Geneva, New York, 1750-1920* (Geneva, N. Y.: Geneva Bicentennial Commission, 1976) 57-78.

59 Stratton, "The Equipages of New York," 46-47.

60 John A. Kouwenhoven, ed., *The New York Guide Book* (New York: Dell Publishing Co., Inc., 1964) 62-63.

61 Henry Collins Brown, ed., *Valentine's Manual of the City of New York for 1916* (New York: The Valentine Company, 1917) 146.

62 Robert J. Vierhile and William J. Vierhile, *The Canandaigua Lake Steamboat Era, 1827 to 1935* (Naples, N. Y.: Naples Historical Society, 1978) 4, back cover. There were other competing steamboat lines, all of them described in this delightful study.

63 *A Streetcar Named Dinky: Canandaigua Street Railways, 1886-1930*, compiled by Marilyn J. Hinkley (Canandaigua, N. Y.: Ontario County Historical Society, 1984) 7-11.

64 *Abstract of the Twelfth Census, 1900* 394-395.

65 Clarence P. Hornung, *The Way It Was: New York 1850-1890* (New York: Schocken Books, 1977) 136.

66 Reginald W. Rives, *The Coaching Club: Its History, Record and Activities* (New York: published privately, 1935) 2, 3, 123, 209, 210, 252, 253.

67 *The Carriage Journal Coaching Number* 3.3.4. (Spring, 1966):144-150.

68 Rives 124, 134, 176, 183, 202.

69 Clarence P. Hornung, *Wheels Across America* (New York: A. S. Barnes & Co., 1959) 59 and Hornung, *The Way It Was: New York 1850-1890* 151.

70 *Abstract of Thirteenth Census, 1910* 505.

71 Joseph J. Schroeder Jr., ed., *1908 Sears, Roebuck Catalogue* (Northfield, Ill.: Digest Books Inc., 1971) 103.

72 *Historical Statistics of the United States, Colonial Times to 1970*, Part 2, 696.

73 *Abstract of the Twelfth Census, 1900* 394-395.

74 Gerald J. Goodwin, Richard N. Current and Paula A. Franklin, *A History of the United States Since 1865*, 2d. ed. (New York: Alfred A. Knopf, Inc., 1985) 433.

75 *Highway Manual of the State of New York* (Albany: James B. Lyon, printer, 1893) 47.

APPENDIX A
The Tax Laws and Tax Records Used in This Essay

1 On April 1, 1799, the New York State legislature levied a tax on real and personal property. The valuation placed on such property was based on figures determined by the United States government's survey of the state. The tax assessors in each town were responsible for three sets of figures: real estate, specifically enumerated personal property and the residue of personal property. The 30 categories of personal property specifically enumerated at predetermined values for tax purposes are listed on Table 1. When completed, the tax assessors' lists were forwarded to the county tax commissioners, who had the authority to correct and revise the sums as well as to adjudicate any appeals. The tax assessors' lists included the names of the owners of the property; the commissioners' lists did not. The commissioners' lists contained the totals for each category of property on a town-by-town basis. It is the commissioners' lists that are available at the New York State Archives in Albany.

In 1799 there were 29 counties in New York State, and the original manuscript tax abstract is available for 24 counties. Abstracts are missing for Albany, Delaware, Kings, Ontario and Rensselaer counties. In 1800 there were 30 counties and abstracts are available for all of them except Dutchess, Oneida, Otsego, Saratoga and Steuben. Thus a record exists for each county in either 1799 or 1800 and, in most cases, a record survives for both years. According to these records, in New York State in 1799 there were 44 coaches, 53 chariots, 18 post chaises, 210 phaetons or coachees, 154 other four-wheel pleasure carriages, 345 two-wheel top carriages and 1,555 other two-wheel pleasure carriages, for a total of 2,379. In 1800 there were 41 coaches, 73 chariots, 13 post chaises, 226 phaetons or coachees, 193 other four-wheel pleasure carriages, 348 two-wheel top carriages and 1,600 other two-wheel pleasure carriages, for a total of 2,494. Since the population of New York State was approximately 588,603 in 1800, less than 1% of the population owned any of the enumerated vehicles. In both 1799 and 1800 there were more enumerated vehicles in New York County than in any other county, even though at this time New York County encompassed only the seven wards at the southern tip of Manhattan Island and the uptown Harlem division.

The geographic pattern described in the text of this essay determined the selection of the counties for analysis–New York, Richmond, Westchester, Albany, Kings, Queens, Suffolk, Clinton and Ontario. The first seven counties are original counties established in 1683. Clinton and Ontario were established in 1788 and 1789, respectively, and represent new settlements. These two counties, the farthest from New York County, underscore the significance of the geographic distribution of the taxed vehicles.

2 On July 2, 1862, Congress enacted comprehensive tax legislation. An annual tax was levied on income in excess of $600. Annual licenses were required on all trades and occupations. Also subject to annual licensing were peddlers, hotel and innkeepers, eating houses, steamers and vessels providing food and lodging, theaters, circuses, bowling alleys and billiard rooms. The licenses cost from $5 to $200. A hotel owner, for example, would have been required to obtain separate licenses for the hotel, for the livery stable and for the retail sale of liquor. He would also have had to pay a tax on billiard rooms.

Monthly taxes were levied on the gross receipts of transportation companies, on the interest on bonds, on surplus funds accumulated by financial institutions and insurance companies, on gross receipts on auction sales and on the sale of slaughtered cattle, hogs and sheep. Gross receipts on newspaper advertisements were taxed quarterly. A stamp tax was placed on all legal and business documents as well as on medicine, playing cards and cosmetics.

Monthly taxes were placed on all manufactured goods, produced by hand or by machine, on the basis of actual sales or the estimated average market value of unsold items. Included in this category were such items as carriages, candles, mineral coals, lard oil, illuminating gas, coal illuminating oil, ground coffee, spices, refined sugar, candy, chocolate, cocoa, starch, tobacco, snuff, cigars, gunpowder, white lead, zinc, pins, paints, clock movements, umbrellas, screws, railroad iron, stoves, paper, soap, salt, pickles, glue, patent leather, sole and rough harness leather, finished or curried upper leather, bend and butt leather, conducting hoses, wine, varnish, furs, cloth, wool, silk, cotton, diamonds and distilled spirits.

Annual taxes were levied on the ownership of carriages, yachts, billiard tables and gold and silver plate. Taxes on carriages were as follows:

$1 tax on a one-horse carriage, gig, chaise, phaeton, wagon, buggy-wagon, carryall, rockaway or other vehicle, the body of which rested on springs of any description, which was kept for use, and not exclusively employed in husbandry or for the transportation of goods, and which was valued at $75 and over including the harness.

$2 tax on two-horse carriages of like description and on any coach, hackney-coach, omnibus, or four-wheel carriage the body of which rested on springs of any description, which was kept for use, or hire, or for passengers, and which was not used exclusively in husbandry or for the transportation of merchandise, and which was valued at $75 and not exceeding $200 including harness.

$5 tax on carriages of like description valued above $200 and not exceeding $600.

$10 tax on carriages of like description valued above $600.

Other personal property taxed:

$5 tax on pleasure or racing vessels, known as yachts, whether by sail or steam, valued under $600.

$10 tax on such vessels valued above $600 and under $1,000.

$10 tax for each additional $1,000 in value of said vessels.

50 cents per troy ounce on plate of gold kept for use.

3 cents per troy ounce on plate of silver kept for use. Exempt from tax were silver spoons or plate of silver not exceeding 40 ounces belonging to any one person.

$10 tax on billiard tables kept for use.

APPENDIX B
Examples of Vehicle Ownership in the 1860s

Because the 1860s tax records still include the names of individual taxpayers, selected individuals were identified as to financial and social status by consulting city directories and county and town histories. These individuals are listed here, along with their vehicles and the amounts at which the vehicles were taxed. The incomes, taken from the tax records, represent the amount reported by the taxpayer above the first $600 of income, which was tax exempt.

NEW YORK COUNTY
The Financial and Commercial District Wards 1, 2 and 4

- T. B. Marsh, occupation unknown, 3 carriages; 1 each at $2, $5 and $10. Income $21,519.70.
- George W. Robbins, occupation unknown, 1 carriage at $1. Income $14,198.20.
- Jacob Philips, pawnbroker, 2 carriages at $1. Income $2,900.
- Joseph A. Morrell(?), physician, 1 carriage at $1. Income $2,135.20.
- J. R. Hoffman, brewer, 1 carriage at $1.
- M. J. de Pumaryo, wholesale dealer, 1 carriage at $1. Income $5,291.
- Rachael Hannan, livery stable keeper, 4 carriages at $2. (The only woman listed for this district.)

City Hall and SoHo Wards 3, 5, 6 and 8
- William Tucker, President of the Knickerbocker Insurance Company, 1 carriage at $1. Income $11,039.
- George P. Trigg, merchant, 2 carriages; 1 at $1 and 1 at $5. Income $8,155.
- George Wilkes, physician, 3 carriages; 1 each at $1, $2 and $5. Income $5,592. Doctor Wilkes was the President of the New York Medical and Surgical Society.
- Joseph B. Taylor, brewer, 2 carriages; 1 at $2 and 1 at $10.

Greenwich Village and Chelsea Wards 9, 15 and 16
Ward 9
- Archibald G. Amour, manufacturer of carriages at 8 Seventh Avenue.
- Edward Senior, undertaker at 75 Carmine and livery stable keeper at 57 Downing, 11 vehicles; 1 at $1, 4 at $2, 6 at $5. Mr. Senior's residence was uptown, at West 88th Street.
- Joseph Falman, occupation unknown, 3 carriages; 1 each at $1, $5 and $10.
- Isaac Bernheimer, merchant, 1 carriage at $10.
- Mrs. Maria Lent, livery stable keeper, 7 carriages; 2 at $2, 5 at $5.
- Maria Gird, 1 carriage at $1.
- Christopher Gwyer, butcher, 3 carriages; 1 each at $1, $2 and $5.
- Jacob R. Reid, butcher, 1 carriage at $1.

For the most part this pattern repeats itself for any number of occupations, trades and professions from grocers to carpenters to physicians. Some individuals in each category owned more than one vehicle, while others owned just one vehicle.

Ward 15
- William B. Astor,* [1]president of the Astor Library, 4 carriages; 1 at $1, 1 at $2, 2 at $5. Also listed 8,400 ounces of silver. William B. Astor was the son of John Jacob Astor.

William's principal concern was the administration of the family's vast interests, especially the Astor Library; he was known as the "landlord of New York" because he continued his father's policy of buying New York City real estate.[2]
- Cornelius V. S. Roosevelt,* Merchant, Director Chemical Bank, grandfather of Theodore Roosevelt, 2 carriages at $5. Income $63,089.96.
- James Roosevelt,* former judge, 3 carriages at $2.
- Silas W. Roosevelt,* lawyer, 1 carriage at $1.
- "Commodore" Cornelius Vanderbilt, merchant, president of Atlantic and Pacific Steamship Company, 7 carriages; 5 at $1, 2 at $2. 512 ounces of silver. One of the Commodore's many ventures was the construction of a combination steamship and railway route to California's gold coast via Nicaragua, which was two days shorter than a competing route through Panama. (See William H. Aspinwall, Westchester County for the competing route.)
- Lloyd Aspinwall, merchant, director of Bank of State of New York, 2 carriages; 1 at $1, 1 at $2. Also listed 1 yacht at $5. The brother and associate of William H. Aspinwall.
- Henry Chauncey, merchant, president of Bergenport Copper Company, 4 carriages; 1 at $1, 1 at $2, 2 at $5. Another associate of William H. Aspinwall.
- Henry Chauncey Jr., merchant, 5 carriages; 1 at $1, 1 at $2, 3 at $5.
- James P. Kernochran, merchant, 4 carriages; 2 at $1, 1 at $5, 1 at $10.
- James Brown,* director of Bank of North America, 5 carriages; 2 at $1, 3 at $5.
- Joseph Sampson,* director of Bank of Commerce, 4 carriages; 1 at $1, 3 at $5.
- Benjamin L. Swan,* director of Bank of America (which had $3 million in assets), 2 carriages at $2.
- Thomas Tileston,* merchant, president and director of Phoenix Bank, chairman of New York Clearing House Association, director of Farmer's Trust Company, trustee of United States Trust Company, 5 carriages; 2 at $1, 1 at $2, 2 at $5.
- B. C. Morris, president of Columbian Insurance Company, 3 carriages; 1 each at $1, $2 and $5.
- David Thompson, president of New York Life Insurance Company, 3 carriages; 1 each at $1, $2 and $5.
- Samuel Lord, dry-goods retailer, 2 carriages at $5. The "Lord" of Lord and Taylor's. (See page 78 for an illustration of one of their emporiums.)
- A. T. Stewart,* dry-goods retailer, director of Merchants Bank, 4 carriages; 2 at $2, 2 at $5. Previously mentioned as the American originator of the department store. In 1862, Stewart surpassed his "Marble Palace" by erecting an eight-story cast-iron retail store that occupied the entire block from 8th to 9th Streets and from Broadway to Fourth Avenue. Construction cost $2,750,000. This store was staffed by 2,000 employees. During the Civil War, Stewart obtained Army and Navy contracts to supply the military with uniforms. In 1869 Stewart established the "model town" of Garden City on Long Island.
- Peter Lorillard,* 5 carriages; 2 at $1,, 3 at $5 and, at his Westchester residence, 3 carriages; 2 at $1, 1 at $5.
- Pierre Lorillard,* 3 carriages; 2 at $1, 1 at $10. The Lorillards were manufacturers and retailers of snuff and the founders of today's P. Lorillard Tobacco Company.
- Willard Parker, surgeon, 3 carriages; 2 at $1, 1 at $5.
- H. J. Raymond, editor, *New York Times*, 1 carriage at $1.
- Eli White, furrier of Eli White and Sons, 2 carriages; 1 at $2, 1 at $5.

*See Note 1 to Appendix B

- James Lenox,* 6 carriages; 1 at $1, 1 at $2, 3 at $5, 1 at $10. Also listed 6,681 ounces silver plate, 3 1/4 ounces gold. Income $83,458.25. Lenox was a realtor who was also a collector of books; his collection included the first Gutenberg Bible in the United States. In 1861 Lenox authored Shakespeare's Plays in Folio and The Early Edition of King James' Bible in Folio. He founded the Lenox Library in 1870. Lenox also donated the land for the Presbyterian Hospital.

- Louisa Lafarge, widow of John,* 1 carriage at $5. Mrs. Lafarge's husband, a Frenchman, had been an officer in both the army and navy during the Napoleonic era. When he left the military, he settled in the United States and became involved in real estate. Their son John was a renowned painter of murals, but his specialty was stained-glass windows. Examples of his work adorn the walls of the Minnesota State Capitol and the windows of several New York City churches.

- William C. Rhinelander,* 3 carriages; 1 at $2, 2 at $5.
- Archibald A. Amour, livery stable, 10 carriages; 3 at $1, 4 at $2 and 3 at $5.
- George Murray, livery stable, 5 carriages at $5.

Ward 16

- Don Alonzo Cushman,* merchant, director of Clinton Insurance Company, founder of Greenwich Savings Bank, 3 carriages; 1 at $1, 2 at $5.
- Clement C. Moore,* author of "A Visit From St. Nicholas" and professor emeritus Oriental and Greek Literature at the General Theological Seminary, 2 carriages at $1.
- Reverend Robert S. Howland, Holy Apostles Church (Episcopal), 6 carriages; 4 at $1, 2 at $5. Also listed 1 yacht at $5 and 929 ounces of silver.
- Henry Van Schaick, lawyer, 1 carriage at $1.
- Edmund H. Miller, broker, 5 carriages; 1 at $1, 1 at $5, 3 at $10.
- William G. Mickel, occupation unknown, 4 carriages; 3 at $5, 1 at $10.
- Mary E. Squire, wife of a dry-goods retailer, 2 carriages; 1 at $1, 1 at $5.
- Catherine Lowerre, wife of a lawyer, 3 carriages; 1 at $1, 2 at $5.
- Knickerbocker Stage Company (capital of $200,000,) 70 carriages; 2 at $1, 11 at $2, 57 at $5.
- Robert Barkley, president of the Knickerbocker Stage Company, director of Citizens Bank, 1 carriage at $1.

From 14th Street to 40th Street (except Chelsea, discussed above) Wards 18, 20 and 21
Ward 18

- Peter Cooper, merchant, trustee of United States Trust Company, 4 carriages; 1 at $1, 2 at $5, 1 at $10. The founder of the Cooper Institute.
- George Griswoll,* merchant, 3 carriages; 2 at $1, 1 at $5. Also listed 2 yachts at $10 and 7,390 ounces of silver.
- Theodore Roosevelt,* merchant, father of President Theodore Roosevelt, 2 carriages; 1 at $1, 1 at $2. Income $37,456. A Gramercy Park resident.
- Ed. S. Brooks, Brooks Brothers Clothing, 5 carriages; 2 at $2, 3 at $5.
- Elisha Brooks, Brooks Brothers Clothing, 4 carriages; 1 at $1, 1 at $2, 2 at $5.
- John Brooks, Brooks Brothers Clothing, 3 carriages; 1 each at $1, $2 and $10. This exclusive men's clothing store was established in 1817 by Henry Sands Brooks at Catherine and Cherry Streets in Ward 4. When Henry died the business passed to his sons. In 1858 a second location was opened at Broadway and Bond Street in Ward 14. By 1861 the firm had 400 employees. During the Civil War the firm obtained a contract from New York State to make uniforms. The Cherry Street store was looted during the draft riots of 1863. Abraham Lincoln and Generals Grant and Sherman were among the illustrious customers of this store.[3]

- Charles G. Gunther of C. G. Gunther & Sons, furriers, 3 carriages; 1 each at $1, $5 and $10. Mayor of New York City in 1865. When Gunther retired from politics he built the Brooklyn, Bath and Coney Island Railroad because he recognized the potential of Coney Island as a recreational area.[4]
- J. C. Gunther, Gunther & Sons, furriers, 3 carriages; 1 each at $1, $2 and $5. Also listed 1 yacht at $5.
- W. H. Gunther, Gunther & Sons, furriers, 3 carriages; 2 at $5, 1 at $10.
- Jacob Lorillard,* 5 carriages; 2 at $1, 1 at $2, 2 at $5. Income $45,129.
- Spencer Lorillard,* 8 carriages; 3 at $1, 2 at $2, 2 at $5, 1 at $10. Also listed 970 ounces of silver. (See Westchester County for the vehicles maintained at his country residence.)
- Samuel Millbank of Millbank & Sons, brewers, 3 carriages; 1 each at $1, $2 and $5.
- Robert Stuart, sugar refiner, 4 carriages; 1 at $2, 2 at $5, 1 at $10.
- Moses Taylor,* merchant, president of City Bank, 4 carriages; 1 at $1, 1 at $2, 2 at $5. A Fifth Avenue resident.
- August Belmont, principal of August Belmont & Co., bankers, director of Bank of State of New York, president of Patriotic Central Aid Association, 6 carriages; 1 at $1, 2 at $2, 3 at $5. Belmont emigrated to the United States in 1837. He was from a wealthy family and had been associated with the Rothschild banking firm. He established a banking company on Wall Street, married the daughter of Commodore Perry and served as minister to the Netherlands from 1853 to 1857. When the Civil War broke out he actively supported the Union cause and pressured his European contacts to aid the North. Active in the Democratic Party, Belmont attended state and national conventions as a delegate and as a presiding officer. A Fifth Avenue resident.
- William G. Read, director of City Bank, 6 carriages; 3 at $1, 3 at $2.
- J. D. Wolfe,* director of City Bank, 7 carriages; 4 at $1, 3 at $5. (See Westchester County for vehicles kept at country residence.)
- Tredwell Ketcham, broker, Tredwell Ketcham & Co., 2 carriages; 1 at $1, 1 at $2. Also listed 1 yacht at $5. Income $100,952.
- Charles Butler, president of Wabash & Erie Canal Company, president of New York Infirmary for Women and Children, 5 carriages; 3 at $1, 1 at $2, 1 at $5.
- Thomas W. Kennerd, chief engineer, Atlantic & Great Western Railroad, 2 carriages; 1 at $2, 1 at $10.
- Sidney Mason, president of Sixth Avenue City Railroad, 4 carriages; 1 at $1, 1 at $2, 2 at $5.
- Samuel Sloan, president of Hudson Railroad, 2 carriages; 1 at $1, 1 at $5.
- George G. Barnard, New York State Supreme Court Judge, 2 carriages; 1 at $1, 1 at $10.
- Valentine Mott,* surgeon, president of New York Medical Society, president of Women's Central Association of Relief for the Army, founder of Bellevue Hospital, 2 carriages; 1 at $1, 1 at $5. Dr. Mott resided at Number 1 Gramercy Park.
- Phillip Phoenix, lawyer, 6 carriages; 2 each at $1, $2 and $5. Also listed 1 yacht at $5.
- Samuel J. Tilden, lawyer and politician, 1 carriage at $1. Governor of New York State in 1875 and the unsuccessful presidential candidate in the disputed 1876 election.

- General Winfield Scott, president of Protective War Claim Association, 1 carriage at $1.
- Abram Wakeman, lawyer, postmaster of New York County, 2 carriages; 1 at $1, 1 at $10.
- R. Stuyvesant, 3 carriages; 1 at $1, 1 at $5, 1 at $10. Also listed 1 yacht at $30. A Fifth Avenue resident.
- Hamilton Fish, lawyer and politician, United States House of Representatives from 1843 to 1845, Governor of New York State in 1848, United States Senate from 1851 to 1853, United States Secretary of State from 1869 to 1877, 5 carriages; 1 at $1, 1 at $2, 3 at $5. During the Civil War he was a member of the Union Defense Committee and a commissioner for the Relief of Prisoners.
- Julia Fish, wife of Hamilton Fish, a member of the Advisory Council of the School of Design for Women at the Cooper Institute, 1 carriage at $1.
- Julia G. Jerome, 2 carriages at $10.
- Mrs. L. R. Jerome, wife of a broker, 3 carriages; 1 each at $2, $5 and $10.
- Julia A. Moffat, wife of a retailer of patent medicine, 3 carriages; 1 each at $1, $2 and $10.
- Caroline Stokes, widow of Henry Stokes, president of Manhattan Insurance Company, 1 carriage at $2.

Ward 20
- George Moore, distiller and wholesale liquor dealer, 3 carriages at $1.
- Robert Ray, vice president and director of Bank of Commerce, 3 carriages; 1 each at $2, $5 and $10.
- F. Fersenheim, occupation unknown, 4 carriages; 1 at $1, 2 at $2, 1 at $5.
- John H. Saffen, occupation unknown, 1 carriage at $10.
- Thomas and John McLelland, owners of a stage line, 42 carriages at $2 and 2 at $5.

Ward 21
- John Jacob Astor, Jr.,* trustee of Republic Insurance Company and of United States Trust Company, 2 carriages at $2. Also listed 1,542 ounces of silver. Income $31,298. A son of John Jacob Astor.
- William Astor,* 3 carriages; 1 each at $1, $2 and $5. Also listed 1,324 ounces of silver. Income $40,632. Grandson of John Jacob Astor and the son of William B. Astor, listed above.
- F. A. Palmer, president of Bank of Broadway and trustee of Broadway Bank for Savings, 5 carriages; 1 at $1, 2 at $2, 1 at $5, 1 at $10.
- Frederick Schuchardt,* director of Bank of New York, 5 carriages; 2 at $1, 2 at $5, 1 at $10.
- George Law, president of Eighth Avenue City Railroad and Ninth Avenue City Railroad, director of Dry Dock Bank, 8 carriages; 3 at $1, 3 at $2, 2 at $5. Also listed 1,708 ounces of silver.
- R. R. Stuyvesant, 12 carriages; 9 at $1, 2 at $2, 1 at $10.
- John Jay, 6 carriages; 2 each at $1, $2 and $5. Also listed 1,010 ounces of silver. Grandson of John Jay, the first Chief Justice of the United States Supreme Court.
- General Charles Yates, 2 carriages; 1 at $1, 1 at $5.
- Charles Tiffany, jeweler, owner of Tiffany and Co., director of Pacific Bank, 1 carriage at $5. Income $63,703. Tiffany, a descendant of one of the first settlers of the Massachusetts Bay Colony, and John B. Young opened a stationery and notions store in 1837 in the City Hall area. Gradually they offered other items for sale, particularly glassware and jewelry. They added another partner and became known as Tiffany, Young and Ellis. During the wave of revolutions that swept across Europe in 1848 the firm bought many fine pieces of jewelry from royal and noble families. The firm also manufactured and designed

jewelry. When Young and Ellis retired, sole ownership passed to Tiffany, who gradually moved the firm uptown to its present location on Fifth Avenue.
- Susannah Alvord,* 1 carriage at $2
- Susan Alvord,* 2 carriages; 1 at $5, 1 at 410.
- Charlotte Brinckerhoff,* 2 carriages at $5.
- Julia M. Schermerhorn,* 2 carriages; 1 at $1,1 at $2.
- Caroline Suydam,* 2 carriages; 1 at $1, 1 at $5.

- Not all of the residents of this ward were as affluent as the individuals indicated above. In comparison, consider the following:
- Richard Edwards, butcher, 1 carriage at $1.
- John Elder, plumber, 1 carriage at $1.
- William Kirby, carman, 1 carriage at $1.
- Edmund F. Rogers, builder, 2 carriages; 1 at $1, 1 at $2. None of these individuals reported owning any other taxable personal property.

Uptown Wards 12, 19 and 22 and Randall's, Blackwell's and Ward's Islands
Fort Washington in Ward 12
- James Gordon Bennett, owner and editor of the *New York Herald*, 9 carriages; 1 at $1, 8 at $2. Also listed 1,441 ounces of silver plate, a billiard table, 1 yacht at $5 and 1 yacht at $50. Obviously Bennett's nose for news had paid off handsomely. When he retired, his son James Gordon Bennett took over the duties of the *Herald* and scored a journalistic coup when his reporter Henry Stanley "found" Dr. David Livingstone in Africa. The son was also a founder of the New York Coaching Club in 1875.
- Charles M. Connolly, owner of the Charles M. Connolly Co., a tobacco firm, 4 carriages; 1 at $2, 2 at $5, 1 at $10. Income $233,405.88.
- Charles O'Connor, occupation unknown, 3 carriages, 2 at $2, 1 at $10. Income $17,411.47.
- George Lewis Jr., occupation unknown, 5 carriages; 2 at $1, 2 at $5, 1 at $10. Income $9,877.55.
- Shepherd Knapp, president of Mechanics Bank, trustee of the United States Trust Company, 3 carriages; 1 each at $1, $2 and $5. Income $6,702.57.
- Thomas Ingraham, occupation unknown, 1 carriage at $1. Income $3,833,67.

Manhattanville in Ward 12
- Edmund Jones, owner of Edmund Jones & Co., a stationer and retail liquor dealer, 4 vehicles; 1 top wagon at $1, 1 rockaway at $1, 1 top buggy at $2, 1 barouche at $2.
- Daniel F. Tieman, manufacturer of paints who grossed $5,173.78 in December 1862, 5 vehicles; 1 top buggy, 1 rockaway and 1 chaise at $1 each, 2 rockaways at $2.
- T. U. (or L. U.) Peters, minister, 1 carriage at $1. Income $2,909.70.
With the exception of Ingraham and Peters, all of the above individuals reported the ownership of silver. In addition, Bennett, Knapp, O'Connor and Jones reported billiard tables.
- By comparison, R. D. Fiedler, who resided in Ward 19, owned 1 rockaway at $1 and reported an income of $213. Alfred Ing, who resided in Ward 22, also had 1 carriage at $1. Mr. Ing sold tripe at the Centre Street Market in Ward 14. Neither reported silver or billiard tables.

WESTCHESTER COUNTY
- William H. Aspinwall, 5 carriages; 3 at $1, 2 at $5. Also listed 1,616 ounces of silver. Income $30,000. In 1848 Aspinwall and several associates founded the Pacific Mail Steamship Company. The discovery of gold in California

ensured the success of this enterprise. Within two years the company's capital was $2 million, realized from transporting gold to the east. In 1856 Aspinwall and several associates secured a virtual monopoly on this lucrative traffic by building the Panama Railroad. The railroad netted $6 million in 1859. When Aspinwall retired in 1856 he was one of the richest men in New York State.[5]

- Peter Hoe, occupation unknown, 4 carriages; 3 at $1, 1 at $10. Also listed 91 ounces of silver.
- E. A. (& A. M.)? Hawkins, occupation unknown, 2 carriages; 1 at $1, 1 at $10. Also listed 791 ounces of silver.
- J. Butler Wright, occupation unknown, 6 carriages; 1 at $1, 3 at $2, 1 at $5, 1 at $10. Also listed 584 ounces of silver.
- Alfred B. Mead, cattle broker, 2 vehicles; 1 wagon at $1, 1 coach at $5. Also listed 55 ounces of silver. Income $3,000.
- E. D. Hunter, occupation unknown, 4 vehicles; 1 spring wagon at $1, 1 top wagon at $2, 1 rockaway at $5, 1 coach at $5. Also listed 1,737 ounces of silver.
- Richard Lathers, occupation unknown, 2 carriages; 1 coach and 1 barouche, both taxed at $5. Also listed 37 ounces of silver.
- Charles Condert, occupation unknown, 1 coach and 1 barouche, both taxed at $2. Also listed 73 ounces of silver and a billiard table.
- Thomas Cuthbert, occupation unknown, 3 vehicles; 1 wagon at $2, 1 carriage at $2, 1 barouche at $2. Income $5,000.
- Newton Carpenter, 4 vehicles; 1 buggy at $1, 1 top wagon at $2, 1 landau at $2, 1 coupé at $5. Also listed 157 ounces of silver.
- Robert Colgate, occupation unknown, 5 vehicles; 1 wagon at $1, 1 phaeton at $1, 1 coupé at $5, 2 carriages at $5. Also listed 227 ounces of silver and a billiard table. Income $51,500.
- Walton Evant, occupation unknown, 3 vehicles; 1 french drag, 1 rockaway, 1 open carriage, all at $5. Also listed 467 ounces of silver and a billiard table. Income $6,284.40.

ALBANY COUNTY

- John McIntyre, retail grocer, 2 vehicles; 1 buggy at $1, 1 rockaway at $2. Income $1,594.22.
- Bradford R. Wood, attorney, 3 vehicles; 1 buggy at $1, 1 rockaway at $1, 1 stanhope at $2. Income $6,919.60.
- Stephen Van Rensselaer, a son of the "old patroon," 6 carriages; 1 at $1, 4 at $2, 1 at $5.
- Joel Rathbone, 5 vehicles; 2 buggies at $1, 2 carriages at $5, 1 carriage at $10. Income $30,100. He was the president of the Albany and Bethlehem Turnpike, director of the Albany Gas and Light Company, president and director of the Albany Exchange Company, director of the New York State Bank, a governor of the Albany City Hospital and a trustee of the Dudley Observatory.
- Erastus Corning, the owner of E. Corning and Co., 3 carriages, each at $5. Income $69,001.56. He was president of the Albany Pier Co., president and director of the New York Central Railroad, president and director of the Hudson River Bridge Co., president and director of the Albany City Bank, president of the Albany City Savings Institute, president and director of the Mutual Insurance Company, a governor of the Albany City Hospital, a regent of the New York State University and a trustee of the New York State Library. Mr. Corning was elected mayor of Albany four times, serving his last term in 1837. He was elected to the New York State Senate in 1842 and to the United States House of Representatives (as a Democrat) in 1857 and 1861. Mr. Corning founded the present-day Corning Ware Company.
- Thomas W. Olcott, 1 carriage at $5. He was the president and director of the Mechanics and Farmers Bank,

president of the Albany Rural Cemetery, director of the Albany and West Stockbridge Railroad Company, vice president of the Albany Dispensary, a governor of the Albany City Hospital and a trustee of the Dudley Observatory.

ONTARIO COUNTY

- William Rockefeller, the father of John D. Rockefeller, 2 carriages; 1 at $1, and 1 at $2.
- Norman Rockefeller, brother of William, 1 carriage at $1.
- Francis Granger, lawyer and politician, 3 carriages; 1 at $2 and 2 at $5. Granger was one of the organizers of the Anti-Masonic Party and of the Whig Party in New York State. He was a member of the New York State Assembly in 1826-28 and 1830-32 and a member of the United States House of Representatives in 1835-37, 1839-41 and 1841-43. Nominated for governor of New York three times, he was defeated each time. Granger was nominated for the office of vice president of the United States in 1836, along with two other candidates. When none of the candidates received a majority of electoral votes, for the first and only time in American history the election of the vice president was decided in the Senate. Granger lost by a vote of 33-16. He was appointed Postmaster General of the United States by President Harrison and served from May to September of 1841 in that position.[6]
- John A. Granger, brother of Francis, 2 carriages; 1 at $1 and 1 at $5.
- General Winfield Scott, 2 carriages; 1 at $1 and 1 at $5.
- E. G. Lapham, lawyer, 2 carriages at $1. He would be elected to the United States Senate in 1880.
- John and Alexander McKechnie, brewers, 3 carriages at $1.
- Chauncey Musselman, farmer, 3 carriages; 1 at $1 and 2 at $2.

NOTES FOR APPENDIX B

1 In Appendix C of his *Riches, Class, and Power Before the Civil War,* (Lexington Mass.: D. C. Heath, 1973), Edward Pessen lists the wealthiest individuals in New York City in 1845. Many of the owners of taxable vehicles also appear on Pessen's list. Throughout Appendix B of this essay, an asterisk identifies individuals or their direct descendants who appear on that list. According to Pessen, William B. Astor's wealth was assessed at over $500,000 and Cornelius V. S. Roosevelt's was assessed at $250,000 and over. All those Pessen listed had wealth assessed at least at $100,000.
2 Biographical information on vehicle owners is drawn from *The Dictionary of American Biography*, Dumas Malone, ed. (New York: Charles Scribner's Sons, 1936) and *Who Was Who In America 1607-1896*, (Chicago: The A. N. Marquis Co., 1963).
3 Tom Mahoney and Leonard Stone, *The Great Merchants*, new enlarged ed. (New York: Harper & Row, 1966) 36-50.
4 James Grant Wilson, ed., *The Memorial History of the City of New-York* (New York: New-York History Company, 1893), vol. 3, 511.
5 William Aspinwall's associates included his brother Lloyd and Henry Chauncey, both of New York County. See section on New York County.
6 James N. Granger, *Lancelot Granger, A Genealogical History* (Hartford, Conn: Case, Lockwood, & Brainard Co., 1893) 301-305.

APPENDIX C
Definitions of Types of Vehicles

Barouche: A four-wheel vehicle in the *coach* family, including undercarriage, lower quarter panels and doors of a coach, with a folding top covering the back seat. Probably of German origin in the early eighteenth century; became popular for town use in the United States by the early nineteenth century.

Buggy: A light, simply-constructed, four-wheel vehicle developed in America, generally with a folding top. This type of vehicle was mass produced by the second quarter of the nineteenth century and, being inexpensive, was distributed widely throughout the United States. Developed in the early nineteenth century, its use continued into the twentieth century.

Caleche: After 1850 this vehicle was called a "barouche." (see above)

Cart: A two-wheel vehicle of numberous designs, from practical utility carts for freight, refuse, etc. to more elegant or sporting types. One of the earliest forms of wheeled transportation, the cart was used continuously throughout the Carriage Era and into the age of the automobile.

Chariot: A sort of half-coach, generally formal in design and use. Superseded by the brougham and coupé. Developed in Europe in the mid-seventeenth century, appearing in the United States about 1700. By the 1890s, it had limited use.

Chariotee: A vernacular type of four-wheel vehicle similar to the extension-top phaeton, with two seats and a folding top. In use from the late eighteenth century to the 1840s.

Coach: A four-wheel enclosed vehicle with vertical pillars supporting a fixed roof. Used as a private or public conveyance. Appearing in the United States in the late seventeenth century, with greatest number in the south, and used until the end of the era, about 1910.

Coachee: A family carriage developed in America, similar to a coach, in use from about 1790 to 1860.

Coupé: A member of the *coach* family with panels cut down longitudinally and latitudinally. This type was developed in France in the eighteenth century and was popular in the United States from the beginning to the end of the nineteenth century.

Hackney Coach: A public conveyance or carriage for hire, first used in the 1600s in England and until 1860 in the United States.

Landau: A coach with a folding top. Developed in Germany in the late sixteenth century; appearing in England about 1750 and in use in the United States from about 1800 to 1900.

Omnibus: A public conveyance intended to carry passengers, generally with a rear door and longitudinal seats. The prototype was developed in Europe during the early part of the nineteenth century. John Stephenson's 1831 omnibus in New York City established its reputation as a mass transportation vehicle. In use until the end of the Carriage Era for urban transport and service between railroad stations and resort hotels or other destinations.

Peddler's Wagon: A commercial wagon used by peddlers, with drawers and compartments for wares. In use from the early eighteenth century until the end of the Carriage Era; in rural areas it continued to be used until about 1920.

Phaeton: A type of owner-driven vehicle, originally characterized by high suspension; introduced about 1735. By the nineteenth century the term could be applied to over a dozen different types of vehicle.

Pleasure Carriage: A light vehicle of a variety of types, used for activities not related to hauling (i.e. work). Late sixteenth century to the end of the era, about 1910.

Post Chaise: A type of chariot, used as a private or mail carrying vehicle, driven by a postilion. In use in the United States from the mid-eighteenth to the mid-nineteenth century.

Rockaway: A four-wheel covered carriage with paneled or curtain sides, derived from the coachee and germantown. Developed in Jamaica, Long Island about 1840, it was a distinctly American type, with various body styles ranging from the simple to the elaborate. Characterized by an extended roof that protected the driver. Popular until the end of the Carriage Era.

Sleigh: A sliding vehicle on bobs or runners, generally used during the winter months. In the United States, in use from the colonial period into the early part of the automobile age.

Spring Wagon: A square box wagon with movable seatboards hung on elliptic springs. A versatile rural vehicle used from about 1860 to about 1920.

Stagecoach/Stagewagon: A coach or wagon used for public transportation in successive stages. Stages were first used in the United States early in the eighteenth century. The numbers of stages and stage lines increased steadily until the railroad displaced them. Stages continued to be used in rural regions during the early 1900s.

Stanhope/Stanhope Gig: A two-wheel vehicle developed by Fitzroy Stanhope about 1815, mounted on cross springs that were mounted on two side springs. It remained a popular vehicle for both personal and pleasure use until the end of the Carriage Era.

Top Carriage: Refers to all types of vehicles with a hood, including most buggies and rockaways.

Wagon: A relatively heavy, box-body vehicle primarily for hauling freight, although the term could describe general transportation vehicles such as the road wagon. Various types used throughout the Carriage Era; some types, particularly trade or freight vehicles, used until about 1930.

THE SPIRIT OF PROGRESS:

Horse-Drawn Transportation on Long Island, 1800-1900

M. Hunt Hessler

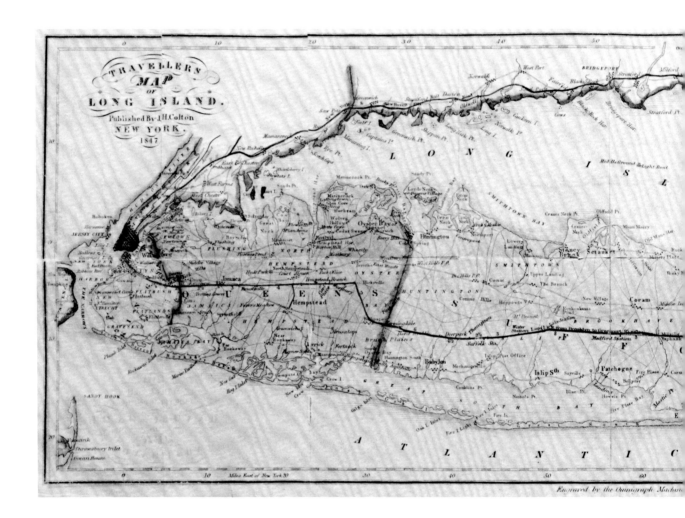

Throughout the nineteenth century, horse-drawn transportation was a central feature of social and economic life on Long Island. As the region grew from its pre-industrial agrarian orientation to the more diversified patterns of life in the later decades of the century, changing patterns of ownership and use of horse-drawn vehicles kept pace; the development and maturation of all forms of conveyance during this period were symptomatic of the island's growth generally. But the story of this process is a complex one; industrialization, the building of the Long Island Railroad, proximity to New York City, social stratification, economic specialization and other agents or reflections of change all contributed to growth trends in varying degrees. Throughout the century, however, one thing remained constant. Despite change in many forms, Long Island retained its rural self-image. The seasonal orientation of this still predominantly agricultural

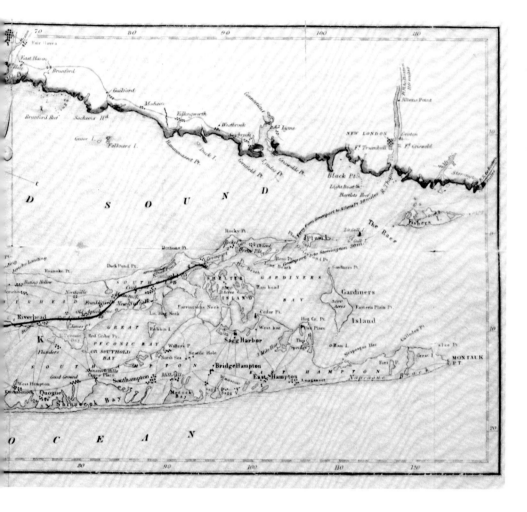

Travellers Map of Long Island
1847
J. H. Colton
New York, New York
Photograph Courtesy of
The Richard H. Handley
Long Island History Room,
The Smithtown Library,
Smithtown, New York

region was not abandoned; while Long Island indeed kept pace with modern forces, it remained in tune with the changing seasons as well.

For the purposes of this essay, Long Island constitutes the area now covered by Queens, Nassau and Suffolk Counties. In the nineteenth century Queens County included all of Nassau, which did not become a separate county until 1899. Unless otherwise noted therefore, the nineteenth-century designation of Queens as the western half and Suffolk as the eastern half of the island will serve as the geographical breakdown for our analysis. Long Island, extending east from New York City for approximately 118 miles, is bounded by water on all sides: the Long Island Sound to the north, the Atlantic Ocean to the east and south and the East River and the Narrows to the west. For much of its earliest history, the most populated areas were usually found either close to the city or along the north and south shores, where many port towns developed to serve the general needs of the region's economy. The 1847 map shown above indicates the shape of the island and many of its population centers after the introduction of the Long Island Railroad, which reached east to the village of Greenport on the Island's North Fork in the year 1844.[1]

THE SPIRIT OF PROGRESS

"The greatest factor in developing any country," C. H. Schmidt wrote in 1890, "is the building of good roads."[2] After witnessing many decades of rapid growth in the region, few Long Islanders would have argued with Schmidt's assessment. Indeed, the centrality of horse-drawn transportation to the local economy had necessitated a constant crusade to upgrade the crude and often hazardous roadways on the island; and, of course, new roads were built whenever possible in the ongoing effort to stimulate local trade and provide increased access to and for the area's many communities. During its early history, however, Long Island had not developed the regional orientation or economic base that would mark its growth throughout the nineteenth century. With a market orientation that looked almost exclusively to the sea, seventeenth- and eighteenth-century Long Island inhabitants did little to connect community to community or eastern villages to those in the west.

The first roads on the island were the village streets of the settlement communities. As these agricultural hamlets spread out, so too did the paths that were cleared to connect the central village to its outlying farms. Generally, though, these "cartways" served the needs of local inhabitants and not the region as a whole. Little was done to bring the neighboring villages together. As Peter Ross noted in his 1903 history of Long Island, the utilitarian value of these first roads–often merely an improvement of old Indian trails–was apparently all that was required:

That the early roads were narrow, crooked and irregular, that they crossed each other in reckless fashion, that they often ended in a farmhouse gate, and a new road has to be discovered or pointed out for further progress, did not seem to detract

from their value in the eyes of the traveler or awaken, apparently, the desire for improvements on the part of the residents. The cow-paths, as they called most of them, were convenient, cost nothing or next to nothing for maintenance or repair, and in a settled community were as good as then could be contrived, or even apparently desired.[3]

In similar fashion, Ralph Henry Gabriel concluded in his 1921 study of the island's "evolution" that these first roadways "represented the simplest adaptations to the primitive environment." He continued:

It must not be believed that these "highways" were, in any way, comparable to the roads of the later time. They were little more than paths set aside for common use. Along them the village herdsman drove the cattle and sheep of the community and men walked to and from their fields. A few carts jolted over them, but rough-hewn sledges were probably more common. Roads which carried such simple traffic needed neither improvements nor uniformity . . . . Some were rented out to local husbandmen and others were kept open for use. These were pioneer days not only in farming and fishing but in road making.[4]

As Gabriel noted, individual residents, using their private funds, often played an important role in the development of the island's roads during this early period when local government resources were reluctantly expended for such purposes. On several occasions, if a farmer needed a road built for transfer of his produce or a miller required an access route to his establishment, he would put up a bond to guarantee maintenance of the proposed cartway and thus ensure that public costs would be kept to a minimum. In 1728, for example, John Hallock Jr. provided a £100 bond to the trustees of Brookhaven to make and maintain a road on "mill-dam" and also a road over "Hill Stream to Setauket . . . . Said highway is to be Sufitient and Conveynient for Carts."[5] Nearly twenty years later, John Tooker took similar action, putting up a bond of £50 in 1746 to make and maintain "a Good Sufficient Cart way" between his property and a farm in Mt. Sinai.[6]

In addition to this localized private enterprise, during the first half of the eighteenth century the provincial government of New York began developing plans for the island's transportation needs on a larger regional scale, something the individual villages had never done. Particularly in light of eastern Long Island's traditional ties to its New England neighbors to the north, the plan to connect Brooklyn to the island's East End was recognized as an important step in unifying the region's maturing communities and growing population. The initial move came in 1704 with plans to connect Brooklyn to East Hampton, but the year 1724, Gabriel noted, "seems to have marked the beginning of a new epoch in Long Island road building."[7] In that year the General Assembly provided plans for a network of roadways to connect the island's many communities and appointed commissioners to oversee the work. In addition to numerous local roads, by 1733 three main "highways" connected east to west across the island: the North, South and Middle Country Roads. But again, the emphasis was on utility and little else:

In places these thoroughfares were left just as nature made them with little done by hand of man except to mark out their boundaries. In modern times they would not be regarded as roads at all, but they fully served their purpose and probably were about as good as the soil permitted or as the sparse population could provide. They were mainly used at first for the transportation of produce and farming supplies, and, supplemented pretty freely, as it seems to us, by cross roads, they served every practical purpose.[8]

Early in the next century, certain Long Islanders joined with the rest of the young nation and gradually began a private campaign of internal improvements. Beginning in 1801 with the incorporation of the Flushing Bridge and Turnpike Company, private capital soon became the primary force in this move to build and improve what became the island's turnpike system. By the 1840s toll roads connected most of the island's communities, either directly or through branches to the North, South and Middle Country Roads. "Wherever the turnpike came," Gabriel noted, "rural isolation began to break down."[9] This form of private enterprise had indeed been necessary, for early-nineteenth-century Long Islanders were still reluctant to contribute to public funding of the roads. As Long Island historian Peter Ross explained, local farmers "struggled with the roads probably twice a year and then thought no more about them." As a result, he continued, there was little call for a regional approach:

Western View of Huntington Village
1841
From Historical Collections of the State of New York
By John W. Barber and Henry Howe
Published by S. Tuttle, New York, New York
Photograph Courtesy of Huntington Historical Society, Huntington, Long Island, New York

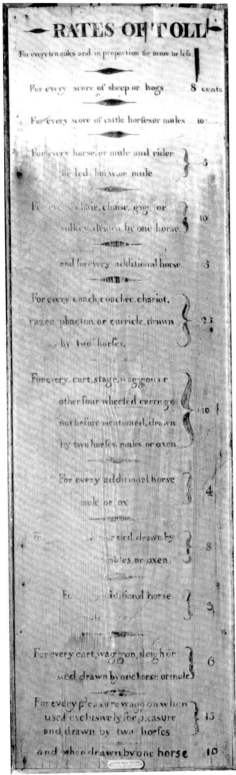

Toll Road Sign
early nineteenth century
Nassau County,
Long Island, New York
Painted wood
Gift of Mr. and Mrs.
G. Gordon Baldwin

The dweller at Southampton did not see he had any business with the condition of the roadway at Islip; those who were supposed by law to look after the roads had no money to effect improvements, and the people, say of Bridgehampton, or Gravesend, would certainly have rebelled had they been assessed for road improvements two miles beyond their limits,—improvements which it was quite probable they might never see, for people did not travel much in those days.[10]

Of course toll roads were not without a price either. Each traveler had to pay for his passage on the roadway, often being charged an amount determined by the type of vehicle making the trip. The more expensive carriages, in effect, would be charged a higher toll. Early on, however, the mere idea that traveling on a road was a privilege to be paid for met with strong opposition. This was, as one commentator put it, "even more obnoxious, as an infringement of natural right, than as a pecuniary tax on the pockets of the people."[11] As the proliferation of toll roads demonstrated, however, farmers soon found that improved road conditions, the time saved, and the decreased wear and tear on their wagons more than compensated for the small charge collected at the toll booth. Indeed, in addition to providing increased access to many parts of the island the turnpike system improved travel conditions on Long Island generally.

Beginning in the 1850s it was hoped that the development of the plank road would solve the recurring problem of muddy and sandy road conditions. For many an enterprising investor, the idea of building a road surface that was uniformly smooth and permanently hard had become, with this innovation, a reality. Nevertheless, their hopes were, in the end, unrealistic. The cost of both wood and general upkeep of these roads proved too expensive for the private investor. In 1857 a Mr. Bailey of New York City wrote to a Smithtown resident who was contemplating investment in a road from Williamsburg to Jamaica, and his estimates of the costs involved were not reassuring. By way of example, he noted that the Saranac Plank Road Company's 22-mile road had an initial cost of $50,000 "to buy the right of way and grade and build the road." In addition, it was estimated that $200 per year per mile was necessary to "keep a Plank road in repair forever." And these costs were already reduced through practical ingenuity: "It is only necessary to Plank one track of a turnpike as in turning out to pass, vehicles very seldom if ever occupy the same track with their wheels, consequently the side of the road which is not planked will not get rutted or cut upon."[12] By 1870, many such ventures were failing. In that year, planks were removed from the turnpike section running between Hempstead and Jamaica, and by the end of that decade the turnpike's entire length no longer had a wood surface.

During this same period the turnpike system itself was also under attack. The *Long Island Farmer* reported in 1869 that "the people of Queens County have been agitating the subject of abolishing toll gates that flood almost every road in the county." There was a time, the report continued, when private investment needed encouragement,

but toll gates have become so numerous that they are positively becoming a drawback to our prosperity, and a nuisance to the increasing multitude of travelers. It will be seen by the call in another place, that the agitations against toll gates and tollbridges are taking shape, and that the people are on the eve of inaugurating a movement that is likely to lead in due time to the abolition on just principles to all concerned, of the numerous franchises that are so inimical to the prosperity of the county.[13]

The anti-toll "movement" in Queens County was indeed taking shape. During the next several years private toll companies were continually brought before the courts because, it was charged, the roads were frequently "almost impassable to vehicles." Very little, however, was accomplished. As the *Long Islander* noted in March of 1875, the situation was bad for both sides. "Exorbitant tolls are charged," it was reported, "and, although no improvements are ever carried on, the corporations are always complaining that they cannot afford to keep the roads up."[14]

This rising tide of opposition resulted in part from an increased move to improve roads through public means. By the third quarter of the nineteenth century, Long Islanders were calling for a system of road building and improvement that would be more responsive to the needs of local communities. Private enterprise could not keep up with the region's growing demands. But tax-supported public management of the roads was not without its drawbacks. Local residents continued to complain about the conditions of their roads, the need for new ones and, of course, the incompetence of those they paid to repair and build them. As the *Long Islander* reported in 1873, "there are but a few duties performed by town or municipal authorities which are more important than making and keeping in repair the common highways."[15] Unfortunately, according to many Long Islanders, this duty was too often neglected and road conditions left in disrepair. The weather was also to blame:

The state of the roads has seldom if ever been worse in this section of Long Island [Huntington] than at the present. The alternate freezing and thawing has reduced them to the condition of a magnified horse radish grater during the present cold spell, and the deep gullies made by the flood of Monday make traveling by wagons in some sections positively dangerous. Most people had rather walk any distance under 20 miles than ride. The wear and tear on wagons is fearful to contemplate. Still it has its compensating effects in that it improves the business of the blacksmith and wheelwright.[16]

Newspaper accounts like this describing the need for road repair and numerous editorials either praising or condemning the actions of local maintenance officials are evidence of the ever-growing importance of horse-drawn transportation to the region's developing economic and social landscape. New road surfaces, new roads, tax-supported maintenance, roadmasters, road commissioners–these were all signs of the increased demands of the late nineteenth century. No longer a region of geographically isolated hamlets, Long Island was now tied together by a network of roads connecting town to town, railroad to seaport, and the island generally to the city and beyond.

Typically, in early-nineteenth-century Long Island a few blacksmiths, a harness maker, a wheelwright, a livery owner and a carriage and wagon maker or two were all that was required to serve a community's farm-oriented transportation needs. Rarely did the artisan work beyond the geographical limits of his own vicinity; the shops were generally small, with little potential for expansion. In fact, according to the census data found in the federal Products of Industry reports and the Reports of Manufacturing, prior to 1840 no carriage maker, wheelwright, harness maker or blacksmith on Long Island accumulated $500 in gross sales, the minimum amount necessary for inclusion in these records.[17] In these categories, and by this official definition, there were no establishments on Long Island that could be classified as "industrial" until the middle decades of the century.

Within the shop economy that dominated artisanal trade during the first decades of the nineteenth century, local craftsmen and those who patronized their services did business on a personal level; face-to-face transactions were the norm. As is typical of such economic arrangements, a barter system often provided an easy method of exchange, a way to buy and sell goods using local commodities. Charles S. Hedges of Sag Harbor, for example, advertised in 1822 that "most kinds of produce (will) be taken in payment" in exchange for a wagon, a carriage or other services rendered. Twenty years later, in 1842, Huntington harness maker William King was more specific. He assured prospective patrons that "Country Produce and Wood (are) taken in payment for Work at the market price." "Grain and Wood" were acceptable at the busy harness and carriage shop of F. G. Sammis. Of all these artisans, though, William Wood and Henry Prime, also of Huntington, were the most direct. Concerned with the rising use of credit–a development traditionally opposed by artisan classes–their 1842 advertisements defined in bold type precisely how they intended to do business: "CASH AND BARTER SYSTEM ONLY." As the island's population grew and the localized economies became more diversified, complex and competitive, this pre-industrial practice was soon abandoned. For many artisans, though, aversion to the credit system did not diminish; in their advertisements, the words "cash only" or "items sold for cash" were frequent reminders of this. Still, faced with increased competition, many carriage makers and skilled craftsmen, such as Hempstead harness maker W. T. Golder, were now

willing to sell their goods "on as reasonable terms as any other establishments."[18]

During the middle decades of the nineteenth century local craftsmen were faced with increased competition from both more ambitious local establishments and producers from the urban market and beyond, and most Long Islanders were not ignorant of this change in their village life. Indeed, some even spoke in nostalgic terms about the days of the "village blacksmith." When Port Jefferson's William Smith, for example, was readying for his retirement to California during the winter months of 1869, it seems that the community felt a personal loss. "Mr. Smith has long been a resident in our village," the editor of the *Long Island Star* reported, "and has been established as the 'village blacksmith' for nearly forty years, and we know of none that will be more missed from our community than he." After the retirement of Riverhead harness maker "Boss" Ketcham, whose "old stand" was both a place of business and a center for local politics, the end of an era was also apparent: "The old harness shop (has) become a thing of the past."[19]

This perception that an age had passed was a constant theme in local Long Island newspapers. The early years of the nineteenth century did not, of course, resemble the idyllic memories born of nostalgic sentiment that some have been led to believe. And yet this perception is quite understandable, for certain dramatic changes occurred during the middle of the nineteenth century that forever altered the face of Long Island life. The most significant of these was the Long Island Railroad. By the year 1844, trains emanating from the westernmost reaches of the island stopped at 13 different locations before arriving at the east end village of Greenport.[20] In the towns along its route, economic activity grew to take advantage of new market opportunities and a new and different network of transportation emerged. Before the introduction of the railroad all farm products and other local commodities not sold locally had to be driven west to distant markets or brought to ships in the various ports. Now, with the train passing through the center of the island for most of its route, each stop was an access point to other parts of the region, to the city and, increasingly, to the national marketplace.

Market wagons were not displaced by this new form of transportation; rather, they became a central feature of the new transportation network that connected the various Long Island farm communities to the regional marketplace at every stop along the rail. The most ambitious effort in this regard–bringing the market wagon and the train together–occurred during the 1880s. At three o'clock on a "gloomy" Monday in January 1885, the "Long Island Railroad was about to try, for the second time, the experiment of carrying farmers' wagons to market by rail." A special train was provided, given the appropriate name of "Farm Wagon Train," containing ten flat cars (for the wagons), ten box cars (for the horses), a passenger coach and a baggage car. "Fifty odd horses" and "huge wagons, loaded with produce, were standing about and the shouting, rugged farmers looked as though they

had just returned from a successful foraging expedition . . . . Twenty two wagons and teams were quickly on board, the farmers crowded into the passenger car and the train moved off." The dramatic difference between the old and new methods for getting to market would escape very few that day:

Crowds gathered along the road to see the train pass. As the farmers on board were rapidly whirled past some brother farmer plodding along in the old fashioned way they derisively cheered him . . . . It takes about seven hours to reach New York from Albertson's by the high road. The train made the run Monday in an hour and thirty minutes. Farmers come from a radius of ten miles around Albertson's and expressed their satisfaction with the new mode of reaching market, which gives them an opportunity to compete with the farmers nearer New York City and Brooklyn.[21]

Despite this initial enthusiasm, however, the "experiment" was short-lived.

Market Wagon
c. 1900
Maker unknown,
United States
Gift of
Mrs. Henry Lewis III,
Charles G. Meyer, Jr.,
G. Howland Meyer and
S. Willets Meyer, in
memory of
G. Howland Leavitt
and his daughter,
Sara Willets Meyer, 1950

This market wagon was originally owned by G. Howland Leavitt, of Bayside, New York.

Produce Being Loaded onto a Long Island Railroad Car
c. 1905
Photograph Courtesy of Suffolk County Historical Society, Riverhead, New York

The $4 fare, many complained, was simply too much. After four weeks only eight wagons made the trip, which was not, it was reported, "a paying affair, and probably will not be until the price is reduced somewhat."[22] "The old fashioned way" of getting to market survived. In fact, the Long Island Railroad, then over 40 years old, had not displaced the market wagon even for trips to the urban market in the west. "One hundred and twenty market wagons, loaded with farm produce, passed our office for New York between two and nine o'clock P.M. yesterday," the *Jamaica Standard* reported in August of 1875, "and it was not a good day for wagons either."[23] Evidently this "traditional" method remained for many the most practical and least expensive method for bringing the island's produce to market.

In certain ways, however, the railroad affected more dramatically the role of horse-drawn transportation on the island itself. Stage lines, for example, responded quickly by altering their routes to fit the schedules of the Long Island Railroad. Gone was the two- or three-day stage ride from Brooklyn to the eastern Long Island shores. In its place were short trips on timetables of exacting detail, arriving at various stations to meet incoming trains. In 1827 a $5 fare would buy passage on a stage that left every other week from Sag Harbor for New York City. Stopping at numerous villages along the way, the passenger was assured of getting "Through in Two Days!" By January 1844, however, this trip across the island was being replaced by shorter and more frequent trips to the closest train station. Mail stages originating in Sag Harbor were now leaving three times each week and traveling west only as far as the "Suffolk Station" (located between Smithtown and Islip), the eastern end of the rail line.[24]

In later years the stage lines became increasingly tied to the exacting schedules of incoming and outgoing trains; Jesse Conklin's "Old Stage Line," between Huntington and Syosset, was a typical example. Announcing his "Winter Arrangement" in January of 1858, Conklin was very precise: "Stages leave Huntington daily at 6 o'clock A.M. to connect with the cars at Syosset, and at 12 o'clock m. to connect with the cars at Hicksville, for Brooklyn. Returning immediately on the arrival of the cars that leave Brooklyn at 10 A.M. and 3:45 P.M." For the increasing number of stage lines taking advantage of the opportunities brought by the railroad, the pattern was consistent. Ezra Seaman's timetable for 1858 was exactly like Conklin's except that it corresponded to the railroad's spring schedules: "The stages will leave Cold Spring for the cars at Syosset, at 6:15 a.m.; returning, leave Syosset on the arrival of the cars that leave Brooklyn at 12 o'clock. Leave Cold

Spring at 3:30 p.m. for Syosset, and return from Syosset on the arrival of the cars that leave Brooklyn at 4:30 p.m." The Queens-to-Brooklyn stage schedules were equally precise. During the winter of 1864 Curtis Powell's "Atlanticville Stage" left Hempstead at 7:45 a.m., arriving in Washington Square at 8:00 a.m., Fosters Meadow at 8.15 a.m. and Brushville at 8:30 a.m. The return trip left Brooklyn, after the arrival of the train, at 1 p.m. the same day. By this period, the train dictated the times and availability of local stagecoaches, as well as their reliability. As the *Long Island Leader* noted in 1875, "as long as the train is on time, the stage is sure to reach its destination . . . no matter how bad the traveling."[25]

The "old fashioned" stage ride was a thing of the past by the middle of the nineteenth century, but the stage itself played a central role in the increased pace of life on the island. Stages were the connecting link between all other forms of transportation. The "Old Medford" line between Port Jefferson and Medford was perhaps the most obvious example, providing intermediate transportation that connected Connecticut to central and eastern Long Island. The route, it was announced in 1875, would continue according to the familiar schedules:

Leaving Port Jefferson at 9 o'clock A.M. to connect with the main train going east; reaching Coram on the way to the station at 10 o'clock.

Farm Wagon Loaded with Cabbages
c. 1905
New Hyde Park, Long Island, New York
Photograph Courtesy of Nassau County Museum New York

*Another stage **Leaves Medford** at 8:2(?) A.M., on the arrival of the Greenport and Sag Harbor branch trains, arriving at Coram 9 o'clock and reaching Port Jefferson at noon, in time to connect with the Steamer Brookhaven, for Bridgeport (Connecticut). All passengers from Bridgeport by the noon boat, wishing to go east or across the Island, will find a stage in readiness to connect them to Medford in time to connect with the eastern bound trains.*[26]

Related trades were also stimulated by the introduction of the railroad and contributed to the island's growing transportation network. Livery stables, blacksmith shops, carriage showrooms and other businesses could now be found situated near the train station where potential profits and competition were high. A. E. Hallock of Smithtown,

for example, chose in 1881 to build his three-story carriage factory—"one of the completest establishments of its kind in this section of the country"—near the railroad station where incoming materials and outgoing vehicles could more easily be transported and potential customers could see his displays. "The large display of carriages," the *Long Islander* reported, "attracts the attention of all passers by." Within two years, Hallock owned "the busiest place in town"; it was, wrote one observer, "a regular hive of industry." After expanding, the new partnership of Hallock and C. B. Vail continued to grow, and the *Long Islander* noted the obvious: "The firm do not intend to confine their business to this vicinity but will traverse all parts of Long Island."[27]

Village of Greenlawn
1880
Edward Lange
(1846-1912)
Long Island, New York
Grisaille watercolor
on paper
Photograph Courtesy of the Charles Tilden Family, Greenlawn-Centerport Historical Association, Greenlawn, New York

Like Hallock and Vail, liveries and related establishments also took advantage of the potential business found near the train stations. In 1884 the *Queens County Sentinel* boasted of several individuals who profited from their proximity to the depot in the busy community of Hempstead:

Our village has three as handsome hacks as will be found in any county village–those of Messrs. Fox, Fagan and Sammis, and visitors here will find their horses can travel as fast as occasion requires. They are always on hand at the depot upon the arrival of the trains. The Livery Stables of Clowes & Hentz, John B. Pettit and Thomas C. Weeks also furnish excellent turnouts for the public.[28]

Clearly, by the third quarter of the nineteenth century many of the carriage-related trades accommodated the changing times. As one might expect, the types of vehicles found on the island's roads changed as well. During the first half of the nineteenth century horse-drawn transportation on Long Island was of a simple variety compared to the diversity that was evident by the century's end, when technological advances, mass production, specialized needs, the ever-growing desire for fashionable vehicles and other developments contributed to the altered face and pace of local life. Although present-day popular perceptions of the period often provide monolithic images of a rural countryside oblivious to change, many nineteenth-century Long Islanders were ever-mindful of the changes around them. Busy village

Station Omnibus
1906
George E. Gould
Lake Grove,
Long Island, New York
Gift of Mrs. Gilbert
Brunelli, 1951

streets, increased trade, railroads and the proliferation of all forms of conveyance were constant reminders to the older generation that the island's former way of life was a thing of the past. In 1875 one such observer noted in not very serious terms how dramatical, and irrepressible, were the agents of change. "Many a man who would have stood still for ever if he could," he wrote, "has been forced to yield to the *spirit of progress.*" As proof, he cited a manifest example:

What a change there has been in Huntington within the last thirty years. It is not so many years ago when our church going people had no more stylish conveyances than a farm wagon, with, perhaps, new sides put on for Sunday. There are men now living who can remember when the spirit of fashion brought out the first so-called carriage . . . . When Squire John brought home his first carriage, and rode to church in it on Sunday, with his pair of fine colts all dressed in new harness, his old fashioned neighbor could not stand it; he must have one too; and so it went. The spirit of change ran through the place, and soon carriages were quite the thing in Huntington, and to be without one was not the thing at all. The fine turnout on Sunday came to be the mark of respectability; and who could afford to be not respectable?[29]

While this reminiscence may not

provide an accurate description of horse-drawn transportation for the earlier period, for many who witnessed the change such comparisons were truisms that reflected a transition from Long Island's exclusively agrarian orientation to one increasingly influenced by the agents of change. For some, a momentary escape from the quickened pace of the late nineteenth century was welcomed. Leisurely carriage rides along the north and south shoes or sojourns to more forgotten areas around the island increasingly became a favorite way to recapture traces of the idyllic past in traditional fashion. The perception, at least, that life was becoming more complicated gave increased importance to the region's natural beauty, still found in abundance beyond the many populated communities, and the traditional use of the carriage took on added significance as both a symbolic and a real link with Long Island's past.

As this form of leisure activity grew in popularity during the final two decades of the nineteenth century, many towns began boasting of their pastoral splendors in order to entice travelers to the vicinity. In 1883, for example, the *Brooklyn Union* spoke in glowing terms of several scenic routes in a piece "About the Beauties of the North Side."[30] Other accounts were more geograph-

**Station Omnibus
at the
Railroad Depot**
*c. 1910
Stony Brook, Long Island,
New York*

*This vehicle was built for
Samuel A. Hawkins of
Stony Brook, and was used
to transport passengers
between the Stony Brook
railroad station and the
nearby village of Setauket.*

(below)
**Driving in the
Shinnecock
Hills of Long Island**
*1892
Lynwood Palmer
(dates unknown)
Oil on canvas
Gift of Lida Bloodgood,
in memory of John R.
Townsend, 1957*

ically specific. One local booster, writing in 1885 under the heading "Smithtown a FAVORED COUNTRY," thought the outskirts of his village would provide a particularly appealing route because "the drive along the north side of the town from Fresh Pond east to the Becar farm . . . is very fine and the drive south along the western side of the Nissequogue river cannot be excelled for beauty of natural scenery. The roads all over the town are broad beautiful drives that tempt one in almost every direction."[31] In Hempstead, "one of the pleasantest drives" ran along Fulton Street, where "the residents to the west . . . made extra efforts to have their gardens look nice."[32] Vehicles were always available for anyone wanting to enjoy these "pleasant and attractive" surroundings:

Those wishing to indulge in an occasional drive should visit the livery stable of Messrs. Hentz & Clowes and examine the many stylish turnouts which that firm have for hire. Those desiring a single seated vehicle and a fast stepper may be readily accommodated, while those who desire to enjoy a family ride in an excellent and easy riding carriage behind horses which apparently know their business, will find equal facilities for being satisfied.[33]

Not coincidentally, the popularity of leisurely carriage drives and the descriptions of the island's rustic beauty reflected the region's growing reputation as a summer resort area. Ironically, then, while some Long Islanders might have bemoaned the passing of simpler times, travelers from nearby New York City came to the island in increasing numbers to escape the pace of urban life. Traditional carriage drives were, of course, part of the lure. That Long Island had never been a model of the English countryside, with elegant carriages forever traveling its roads, seemed of secondary importance to the expectant city traveler. As one local Huntington resident put it, "to those who are now sweltering in the cities a month along these drives would give them a new life through the labors of a busy Fall and Winter."[34] For many potential resort

towns–particularly those connected to the city by rail along the island's south shore–the nostalgic imagery evoked by descriptions of rides in the country was an effective lure for attracting vacationers to the area. In Babylon, for instance, it was advertised that a visitor could expect a change of pace immediately upon arrival "at the familiar lawn-encircled station" where he would "see the same long lines of handsome carriages or capacious teams drawn up in waiting as of old." A similar sight could be found in nearby Bayshore, where a traveler would "be sure to find a large cavalcade of carriages, hacks, and equipages of finer sort drawn up awaiting the arrival of the train." In addition, as one might expect, the area's picturesque carriage drives were always a primary attraction:

The drives of Islip and vicinity also deserve particular mention. A point to which we have already referred is the abundance of fine shade trees, reminding one more of an old New England town than of an island community. This is one of the best fruits of age. Whether one drives east or west, to the north or southward to the shore, the roads will be found firm and level, lined with handsome residences and with many beautiful views constantly presenting themselves. The shore drives are, of course, the most popular. Those to Bay Shore and to Sayville are replete with charming glimpses of the sea. At particular times of the day, especially in the afternoon, a constant stream of elegant carriages flows along these favorite drives.[35]

Throughout the nineteenth century, then, certain aspects of horse-drawn transportation took on the characteristics of the changing economic orientation and faster pace of life found in many Long Island communities. Others, on the other hand, such as the carriage ride, were additions that grew in popularity in part because of these very changes. In sum, as increased opportunities came to the island, a more diverse and complex network of transportation developed accordingly.

One very important activity, however, did not change so dramatically: the sleigh ride. Necessarily, the use of stage sleighs or other business sleighs did change with the economy, but generally the seasonal slowdown of winter remained a central feature of life in this still agricultural region. As a result, sleigh rides retained their traditional place in the often celebrative life of

the winter months. Indeed, nineteenth-century newspapers were quick to note the festive nature of winter sleigh riding. The Union School, for example, organized an annual trek to Oyster Bay on the snow-covered streets.[36] Most often, such outings were village affairs; in February 1874 the Union School Sleigh Ride included 107 sleighs, which understandably attracted considerable attention: bugle calls and church bells, banners and cheers, would announce their entry or departure as they rode through neighboring towns.[37] School children, of course, were not alone in their collective enjoyment of this seasonal pastime. It was reported, for instance, that "150 ladies and gentlemen" rode from Greenport to Riverhead "tooting" their horns on a cold February evening in 1874.[38] As the *Queens County Sentinel* reported, local businesses could also contribute to the winter festivities. If there were any who "wish to indulge in a sleigh-ride," it was announced, "they will find excellent accommodations at Whaley & Son's [Hempstead], not only in the line of refreshments for which the Whaley's are famous, but Prof. Lafricain's services have been secured and he will furnish some of his excellent music for those who desire to dance."[39]

For many, sleigh rides and similar activities remained popular in part because they did not necessarily require a substantial investment. "Even the old crockery crate on bent hickory poles" could provide sufficient means of transportation if, we can assume, a hill and not a horse was the intended source of power.[40] But snow-covered streets also necessitated the need for stage sleighs, business sleighs and other work-related vehicles. The purchase of a new stage sleigh, in fact, was often reported as a significant event. Huntington's stage driver, "Uncle Jesse," for example, bought a locally manufactured sleigh that "comfortably seated" 25 passengers.[41] In this area, local carriage makers were always busy keeping up with demand. In all, sleigh riding, whether for basic transportation or fun, was part of the fabric of social life on Long Island in a way that reflects the seasonal orientation that marked the times. One can almost read the changing tone of the newspaper accounts as they describe the altered pace of winter

months, a time between the more eventful seasons of Spring and Fall.

PATTERNS OF OWNERSHIP

On Long Island, and in the nation generally, the nineteenth century was an extended period of continuous growth and development in almost every aspect of life. In what is now Nassau County, the population grew from 11,102 in 1800 to 26,963 in 1860 and to 55,488 in 1900. To the east, in Suffolk County, population figures were equally impressive: 19,494, 43,275 and 77,582, respectively. In similar fashion, as the population grew and the commercial economy developed, the number and diversity of horse-drawn vehicles increased accordingly. Through a comparative study of these trends the connection between socio-economic development and patterns of ownership becomes clear.

New York State tax records for the year 1800 provide an initial context for this analysis.[42] Included among the enumerated items of personal property in these documents were several vehicle types with specifically designated values: coach ($800), chariot ($700), post chaise ($700), phaeton or coachee with steel springs ($300), other four-wheel pleasure carriages or two-wheel top carriages ($100) and other two-wheel pleasure carriages ($50). The purpose was to single out for taxation those items designated as "pleasure vehicles"; work-related forms of transportation such as farm or market wagons were not considered, nor were sleighs. We can assume, therefore, that the category "pleasure vehicles" also broadly represents vehicle ownership among those most able to purchase such items.

The most striking feature found in the data is simply the number of vehicles listed for Long Island. Queens County had a total of 429 vehicles while, in the east, Suffolk County had 258. Put another way, in these two counties 2.5% and 1.3% of the population, respectively, owned some type of pleasure vehicle.[43] The difference is also striking: The percentage of people owning a pleasure carriage in Queens is almost double that in Suffolk. To broaden the analysis, if we consider the area closest to the city,

William Sidney Mount
and others
in a Sleigh
late 1850s
R. A. Lewis
New York, New York
Gift of Mrs. Robert
Smith (Jae Thompson
Smith), in memory of
Miss Etta Sherry, 1980

Kings County (now Brooklyn), this pattern becomes more distinct. With a population of 5,740 in 1800, Kings County was listed as having 251 vehicles, or 4.4% ownership. Considering the data for all three counties, then, it is clear that proximity to New York City dramatically increased the percentage of ownership.

If ownership of these vehicles can in fact be used as an indication of wealth or social orientation, this geographical pattern should remain consistent. Indeed, a further breakdown of the data indicates differences in the types of vehicles owned in the three counties as well: the more expensive pleasure carriages become more common closer to the city. The following table illustrates this by reducing the data to four categories based on the designated values of the enumerated carriages. Under each county is listed the number and percentage of each vehicle for the area. By way of example, the two Kings County carriages

listed at $800/$700 represents .8% of the total of 251 vehicles located in that county:

VEHICLE VALUE	COUNTY (TOTAL VEHICLES)		
	Kings (251)	Queens (429)	Suffolk (258)
$800/$700	2 (0.8%)	3 (0.7%)	0 (0.0%)
$300	23 (9.2%)	14 (3.3%)	1 (0.4%)
$100	43 (17.1%)	72 (16.8%)	16 (6.2%)
$50	183 (72.9%)	340 (79.3%)	241 (93.4%)

Only a very few inhabitants in the western counties owned the most expensive vehicles: There were a single coach ($800) and two chariots ($700) in Queens and two chariots in Kings; in Suffolk, the eastern half of the island, there were none. Such small numbers in and of themselves do not indicate a trend, but they are consistent with the other data. Taken collectively, each category for each county reinforces the established geographical pattern. In the two more expensive categories the ownership percentage decreases with each move east away from the city; the decrease slows for the less expensive $100 carriages. Conversely, percentages for the least expensive $50 carriages increase in the eastern areas, especially in Suffolk, where 93% of all pleasure vehicles were inexpensive two-wheel carriages. This pattern is best illustrated by rearranging the data to indicate the average value of an enumerated vehicle for each county: $86.65 in Kings, $71.33 in Queens and $54.07 in Suffolk.

These figures should not, of course, be taken out of context. Like Suffolk, Queens County in 1800 was predominantly agricultural–a patchwork of farms, village harbors and only a few more cosmopolitan towns. But within this framework a geographical pattern did exist. In the comparatively isolated hamlets of eastern Long Island, most inhabitants looked to the sea for their main sources of transportation. There was no need, and there were very few passable roads, for the more expensive pleasure carriages found closer to the cosmopolitan center of New York City. Westward, increased access to the urban marketplace, better roads and more concentrated popula-

tion centers gave expensive vehicles a limited place in those areas where society was becoming more stratified and diverse.

By the 1860s much had changed and much remained the same. The "spirit of progress" had altered ownership patterns within Queens and Suffolk Counties themselves, but the island-wide pattern was still intact. Using Federal Internal Revenue tax records for the year 1862, vehicle distribution data for a comparative chronological analysis is available.[44] Once again, pleasure carriages were singled out as luxury items for tax purposes. And, once again, the enumerated carriages can be divided into four groups according to predetermined values: one-horse, spring wagon/carriage valued at $75 or over, with harness ($1 tax); two-horse spring wagon/carriage, or four-wheel vehicle, valued at $75 to $200, with harness ($2 tax); carriages valued between $200 and $600 ($5 tax); and, finally, carriages worth more than $600 ($10 tax). Moreover, because the information was usually recorded for each separate community, the geographical distribution of ownership can be documented in substantial detail, from village to village,

Main Street,
Looking West
c. 1900
Huntington,
Long Island, New York
Photograph Courtesy of
The Long Island Collection,
The Queens Borough
Public Library
Jamaica, New York

township to township, and county to county.

Beginning with the total number of pleasure carriages of Suffolk and Queens, the east-west pattern is again evident. According to the records, Suffolk's 43,275 inhabitants owned 530 vehicles (1.2% ownership); Queens County's population of 57,391 owned 1,317 (2.3% ownership).[45] These percentages are remarkably consistent with the 1800 figures of 1.3% and 2.5% respectively. To pinpoint this pattern further, if we divide both counties into eastern and western halves, lower ownership percentages are indicated with each move east: 2.5% in western Queens, today's Queens County; 2.1% in eastern Queens, now Nassau County; 1.6% in western Suffolk; and, finally, 1.2% in eastern Suffolk. In this area, at least, the geographical pattern of ownership remained fairly constant.

Extending this analysis even further, samplings of township and village ownership patterns consistently support the conclusion that vehicle distribution reflects the socioeconomic orientation of the region. For example, in Suffolk's most rural townships, ownership percentages are the lowest: 0.8% in Shelter Island, 0.9% in Southold, 1.1% in Southampton and 1.2% in East Hampton. But even within these areas the distribution of vehicles is differentiated according to the social development and economic centrality of particular villages. In Southampton Township, for instance, over 50% of all pleasure vehicles listed, 38 out of 74, were located in Sag Harbor, the area's economic center and by far Suffolk's most populated village.[46] Similarly, in Islip Township, a less populated area located on central Suffolk's south shore, over 65% of the vehicles, 30 of 46—and all six of the area's carriages valued over $200—were owned by residents of Islip village, the only community in the township with direct access to the Long Island Railroad.[47]

Although the general geographical distribution of pleasure carriages for the year 1862 was consistent with the 1800 trend, the types of vehicles found in each area no longer fit the established pattern. To illustrate, the following table lists vehicle distribution in western Queens (listed as Queens), eastern Queens (Nassau County) and Suffolk:

VEHICLE VALUE	COUNTY (TOTAL VEHICLES)		
	Queens (752)	*Nassau (565)*	*Suffolk (530)*
over $600	0 (0.0%)	0 (0.0%)	0 (0.0%)
$200–$600	87 (11.6%)	15 (2.7%)	9 (1.7%)
$75–$200	221 (29.4%)	220 (38.9%)	190 (35.9%)
$75 or over (1-Hor)	444 (59.0%)	330 (58.4%)	331 (62.5%)

First, and most obviously, according to the tax records not a single Long Islander owned a pleasure carriage valued over $600. The more elaborate vehicles, it would appear, were not part of Long Island's social landscape. Carriages ranging in price from $200 to $600, however, were found with increasing frequency closer to the city. The geographical distribution of these vehicles is consistent with the general pattern discussed above; in the two remaining categories the pattern is less distinct. Ownership percentages for two-horse vehicles valued between $75 and $200 were lower in central and eastern Long Island than in what is now Queens, but outside Queens, they were nearly equal: 38.9% (Nassau) and 35.9% (Suffolk). In addition, percentage levels for the last category, one-horse vehicles valued at $75 and over, are relatively constant for all three areas: 59.0% (Queens), 58.4% (Nassau) and 62.5% (Suffolk). Considering the size of the sample identified in the tax documents, the differences between these figures are statistically insignificant. Here again the pattern of ownership reflects the changing social and economic fabric of Long Island society. With the help of the Long Island Railroad and new and better roads, in addition to many other factors, economic and social life on the island were becoming more standardized. Eastern Long Island was no longer a region of geographically isolated communities, and its socioeconomic orientation increasingly resembled that of its neighboring county to the west. Horse-drawn transportation, a central and identifying feature of life on the island, inevitably reflected this change.

Ownership patterns were also profoundly influenced by the growing diversity of horse-drawn transportation throughout the century. Large-scale production, specialized designs and technological improvements all contributed to a proliferation of new and different types of vehicles. This phenomenon can be isolated for analysis by examining probate inventories from a specifically designated area. To assess inheritance taxes, New York state required that a decedent's estate be inventoried, listing all real and personal property, which included all carriages, wagons, carts, sleighs and other forms of transportation. Five hundred Queens County inventories are considered here for analysis, one hundred each for the years 1815–16, 1837–38, 1862–63, 1882–83 and 1897–98.[48] Unfortunately, these records do not include an accurate cross section of society. Rather, they are collectively representative of the middle and upper classes– those with estates substantial enough for official consideration. Consequently, they also include disproportionately those most likely to own some form of private transportation. As would be expected, the inventories also list every item from the estate, no matter how old or insignificant. As a result of these two factors, individual ownership figures are quite high. For our purposes, though, this data isolates the dramatic changes in the types of vehicles owned by those who were most likely to own vehicles. The data is tabulated in the appendix (Table 1) for easy reference and arranged to present visually the growing diversity of horse-drawn transportation throughout the nineteenth century.[49]

The first area worth noting is the number of vehicles (all types) found in each period: 166 in 1815–16, 231 in 1837–38, 355 in 1862–63, 221 in 1881–82, and 217 in 1897–98. This, in turn, corresponds to an average of 1.7, 2.3, 3.6, 2.2 and 2.2 vehicles per individual, respectively, for each period. These totals indicate a steady and yet dramatic rise in the number of vehicles owned in the several decades prior to the Civil War. After the war, the totals drop significantly and remain constant for the remainder of the century.

One factor contributing to the increase during the first half of the century is clear. The types of vehicles contributing most significantly to the rise in ownership prior to the Civil War were different from those owned in previous years.[50] For the period 1815–16 there were only 16 vehicle types listed in the inventories. The number of vehicles of these types owned in each of the pre-Civil War periods is remarkably constant;

166 in 1815–16; 167 in 1837–38; 169 in 1862–63. Similarly, if we take all categories listed for 1837–38 and compare their totals to those of the same categories in 1862–63, they are again almost equivalent: 231 and 228, respectively. For each of these periods increased ownership resulted from ownership of new vehicle types not found in preceding years. In effect, this illustrates dramatically how the proliferation and growing sophistication of horse-drawn transportation are two sides of the same coin. New and specialized vehicles did not displace "traditional" ones; rather, they provided a more diversified and specialized network of vehicles for use in the developing economic and social life of Queens County, Long Island.

Still, to qualify this conclusion it should be noted that within the category loosely called "traditional" there is also change. Although the totals are relatively equivalent for the three periods, ownership patterns (or descriptions of them) are not constant. The most obvious example is found with the categories "wagon" and "farm wagon." Ownership of the former decreased as totals for the latter increased, dramatically; in 1815–16, 81 vehicles are listed as "wagons" while only one is described as a "farm wagon;" in 1837–38, only 7 vehicles are listed as "wagons," while the number of "farm wagons" has risen to 29; by 1862–63, 29 "wagons" and 66 "farm wagons" are listed.

Obviously, much of this can be explained by the way the information was originally recorded. Many of the 81 wagons listed for 1815–16, if recorded in 1837–38 would doubtless have been listed as farm wagons or other vehicle types. Indeed, data for all periods suggest that along with the proliferation of vehicle types–and perhaps as a cause of this proliferation as well–there developed in this age of the "transportation revolution" a more distinct awareness of the specialties attributable to all forms of transportation generally. By the end of the century, the inventories provide us with a taxonomy of horse-drawn vehicles that bears little resemblance to early-nineteenth-century classifications. Carriages, wagons, sleighs and sleds are increasingly classified according to construction detail, use, the number of seats or horses or other identifying characteristics. Identical vehicles are described several different ways, each description dependent upon the level of detail the recorder deemed appropriate. In all, 157 categories are included by the end of the century. For those who accumulated this data, the proliferation and growing diversity of horse-drawn transportation necessitated descriptive distinctions of increasing detail; as their efforts document quite clearly, times had changed.

Taken collectively, the data bring into clear focus another significant factor: legally, comparatively few women owned their own vehicles. In fact, few women–only 18.4% of the 500–were included in the inventories. Although women were better represented as the century progressed they never accounted for more than 28% of those inventoried:

	MEN	WOMEN
1815–6	86	14
1837–8	89	11
1862–3	84	16
1881–2	77	23
1897–8	72	28

Telling as these figures are, economic and legal status of women is further indicated by the limited number of those owning a vehicle or a horse. Of the 92 women included in the 500 inventories, only 16 owned some form of transportation, including horses. Of these, seven had no horse to pull their carriage or wagon. Throughout the century, women collectively averaged 0.45 vehicles per individual, while the lowest figure of ownership for both men and women in a given period, 1815–16, was almost four times greater–1.7. Even this comparison understates the case. First, several wealthy women had abnormally large numbers of vehicles, thus inflating the overall figure to 0.45. Secondly, the 1815–16 figure of 1.7 includes 14 women none of whom owned a single vehicle or horse; consequently, the average ownership by men would be considerably higher than the 1.7 overall average.

Throughout the nineteenth century, at least in Queens County, horse-drawn transportation was apparently a male domain. Most likely, if a husband died first his estate would be inventoried and all vehicles and horses would then go to his male heirs, leaving the wife (and later her inventoried estate) only domestic items. If the wife died first, her estate would include neither horse nor vehicle because both were legally and, in the view of society, rightfully the husband's property. Only the anomalous situation would place these items in a woman's estate and, as indicated above, such situations were sometimes the outgrowth of considerable wealth and social status. Maria Ann Linnington of Jamaica (1887), for example, had an estate valued at $52,353.08, complete with a wagon shed and carriage house containing the following:

(2) Top depot wagon, Spring wagon,
Top buggy,
Farm wagon,
Cart Gray, farm team,
(2) Horses Phaeton,
Sleigh, One-horse farm wagon

The remaining 15 estates suggest, however, that wealth was not the primary reason for female ownership. As already mentioned, seven of these estates included no horses, and several vehicles listed were either old wagons of little value or work-related vehicles. Expensive carriages were few. One could speculate that these few estates represent in part the number of families without male heirs. Whatever the reason, the gender-specific pattern of ownership is very clear. Few women owned vehicles or other means of transportation. That social and legal convention critically disadvantaged women in this area seems obvious. Considering the economic importance of horse-drawn transportation, such a disadvantage was a formidable barrier to financial and social independence.

Female ownership patterns also partly explain why the totals for 1881–82 and 1897–98 are comparatively low. In these years more women were included in the inventories and their low ownership ratio consequently reduced the overall totals. But this does not explain all. It is a familiar story that throughout the nineteenth century industrial development, population growth and other factors decreased the percentage of the work force found on the farm. This development would systematically reduce

Haycutting
c. 1885
Photographer,
Rachel Hicks
Westbury, Long Island,
New York
Photograph Courtesy of
Seaman Collection,
Nassau County Museum,
New York

the ownership ratio of farm vehicles per individual in a given population sample. Also, the impact of these same forces created a larger class of economically disenfranchised men and women who could not afford the luxury of owning their own means of transportation. This in turn indicates another change in ownership of pleasure carriages not reflected in the tax records discussed above. As these documents indicate, the overall ownership percentages did not change significantly from 1800 to 1862 in Queens and Suffolk Counties. Consequently, if, as the probate records demonstrate, individual ownership among those most likely to own a pleasure vehicle was on the rise during the first half of the nineteenth century, then a growing number of individuals on the lower rung of the economic ladder did without. Here, then, is a clear indication of the increasing social stratification of Long Island society, a phenomenon symptomatic of the socioeconomic changes that defined nineteenth-century Long Island life.

The probate documents also provide a means for examining the holdings and occupational diversity of farmers, blacksmiths, and carriage makers. For example, the inventory for Elbert Bogart (North Hempstead, 1897) includes the contents of his blacksmith shop, carriage house and farm. Although his estate, which was worth $47,050.54, was perhaps not typical, the inventory of his property was one of the most extensive and detailed listings of vehicle types:

Platform spring wagon, (2) One-horse carts,
(4) Old farm wagon, (2) Dump cart,
(2) Wood sleds, Two-horse spring wagon,
Old pleasure sleigh, One-horse spring wagon,
Top wagon, (2) Road cart,
(2) Old sleigh,
Spring wagon,
Buckboard wagon, Farm wagon,
Road wagon, (9) Horse, One-horse top wagon,
Buggy wagon

For comparison, John Caughlin, also from North Hempstead (1898) had an estate worth only $4,290.00, but his list of vehicle-related items is also long:

(9) Horse, (3) Dump cart,
Road cart,
(2) Hay wagon,
(2) Box wagon, Bob sled,
Buggy, (3) Farm wagon,
Skeleton wagon,
(2) Market wagon,
Sleigh

As these examples illustrate, individuals who owned blacksmith tools, carriage tools and implements of related trades also owned farm vehicles, farming implements and various livestock. If successful enough to be included in these selective inventories it seems that carriage makers and blacksmiths, harness makers and livery owners were also financially able to own and operate a farm. The lure of agriculture and the seasonal orientation of the economy generally prompted many carriage makers and related craftsmen to supplement their often tenuous incomes with profits from sales of farm commodities. Others were more ambitious. For example, Riverhead's Charles Blydenburgh, eastern Suffolk County's most industrious carriage manufacturer in the 1880s, turned to cranberry farming with considerable success. "With an eye to agricultural thrift," Blydenburgh purchased some "worthless swamp land" adjacent to the railroad tracks in 1884 and persuaded the railroad to build a culvert with a flood gate to drain the area. That done, he now owned "fifty acres of as fine cranberry lands as could be found in the country." Using profits from his growing carriage manufactory, Blydenburgh funded his agricultural interest:

The lands have been put in condition as fast as the owner's limited means would permit and there is now a fine five acre cranberry patch to be seen here. Over $300 worth of berries now lie in Mr. Blydenburgh's shop awaiting shipment and they are splendid specimens of this fruit. This enterprise is in its infancy yet, there being 45 acres more.[51]

Obviously, Blydenburgh's spirited enterprise was atypical, but his interest in agricultural pursuits was not. In the probate records, almost every estate containing artisan supplies and tools also included farm-related vehicles, old and new. As Elbert Bogart's inventory indicates, the list could be quite extensive.

The probate records do not, however, tell the whole story. In this case, as well as many others, the records lack sufficient detail to identify definitively those vehicles that were purely for pleasure and those that were also work-related. Clearly, with these records alone, ownership patterns are difficult to reconstruct. For the most part, tax records from the period primarily identified only pleasure vehicles, stages, coaches, carryalls and similar types. Farm wagons, market wagons and numerous other forms of horse-drawn transportation were tax exempt and thus excluded from the records. Probate inventories, on the other hand, present the opposite problem. Every vehicle, regardless of worth, was included: old carriages no longer in use, broken and discarded wagons and sleighs and sleds of every conceivable description, as well as ox carts, donkey carts and numerous other types that were recorded with insufficient description for accurate classification. Many listings therefore included several generations of vehicles: those long ago retired and those still in use were equally represented. The difficulty, then, is to differentiate between the two. Fortunately, through a comparison between the tax records and the itemized inventories, these problems can be partly resolved.

First, when comparing vehicle ownership of men and women listed in both bodies of evidence the accuracy of the tax records is reinforced. Of the 70 inventories for the year 1863 that were used for comparison, only nine individuals were listed in the 1862 tax records that documented ownership in the year prior to their passing. One might assume, therefore, that approximately 13% of those individuals included in the already upwardly biased probate documents owned pleasure (or other) vehicles singled out for taxation. So, even within this small sample group, one that collectively represents the middle class and more wealthy segments of Queens County, there is a marked separation between those who owned pleasure vehicles and those who did not. Upon closer inspection of the itemized inventories, however, there are a few vehicles, but only a few, that seemingly escaped the tax auditor's eye. John Debevorse of Newtown, for example, owned a "new carriage" valued at $150, R. Duryea's estate (Oyster Bay) included a "buggy wagon" valued at $75 and Katherine Napier's inventory (Jamaica) included a "one horse top wagon" valued at $50. These are, it should be stressed, the exceptions–to a remarkable degree, the inventories generally support the accuracy of the tax records. In fact, one might reasonably argue that Debevorse's "new carriage" could have been purchased after tax information was collected. All others, including the two listed above, were valued at or lower than the minimum value of $75 required of such enumerated items in the tax law. Specifically, Duryea's $75 wagon is valued at the minimum amount while Napier's wagon falls well below the minimum at a mere $50. Thus, we can assume that these vehicles were most likely placed outside the auditor's or owner's consideration. All additional entries in the inventories that could be placed within the category "pleasure vehicle" are of considerably lower value as well, some as little as $10 to $20. Although the probate records consistently undervalue these items, inventoried vehicles not found in the tax lists must still be considered of comparatively little or moderate worth. Many of these are designated as "old" carriages, and those that are not probably could have been.

Although the accuracy of the tax records is supported through this comparison, certain weaknesses are evident as well. Specifically, because farm vehicles and other varied forms of horse-drawn transportation were excluded from the tax records, the full complexity of ownership patterns is hidden from view. As the probate records demonstrate, most individuals owning a pleasure carriage singled out for taxation also owned several other vehicles of various descriptions. Jacob Frost of Northside (North Hempstead), for example, was assessed a $2 tax for a two-horse wagon valued at $100. And yet his estate, amounting to the considerable sum of $34,239.95, contained a modest variety of other wagons, carriages and sleighs: one market wagon ($25), one (possibly very old) carriage ($10), three spring wagons (total value $50), one wood sleigh (($10), two pleasure sleighs (total value $20) and the taxed wagon, now undervalued at $75. Similarly, Daniel Smith's estate (Jamaica South), although far smaller than Frost's at $1,112.37, also included a modest collection. In addition to his one-horse carriage, which was valued at $100 in the tax records, Smith's inventory listed a market wagon ($80), one (possibly old or hand-made) farm wagon ($8), two sleighs ($15 and $5) and a top wagon ($25). And, once again, his pleasure vehicle was now undervalued at

$75. The pattern remains relatively consistent for those listed in both sets of documents: a single pleasure vehicle in addition to four to seven others was the norm, some of the latter obviously quite old and of little use and the remainder of more utilitarian value. Newtown's James Moore owned a one-horse pleasure carriage ($75), a market wagon ($35) and three other vehicles; Great Neck's Henry Allen was also taxed for his one-horse wagon (valued at $150) and had four farm-related wagons of values ranging from $10 to $60. Mary Ann Gustine (Newtown), a woman of considerable wealth, was the only outstanding exception. Gustine's large estate of $47,810.60 included only a single farm wagon ($25); her two-horse pleasure carriage, valued at $300 and assessed a $5 tax in 1862, was absent from the inventory taken less than one year later. Although no conclusions can be drawn from a single example, Gustine's case tends to reinforce the argument concerning gender-differentiated ownership patterns. Men of her economic standing almost without exception retained ownership of pleasure and farm vehicles–old and new–to the end; Mary Ann Gustine's inventory thus remains suggestive, if nothing more. For most males with similar holdings, use and perpetual ownership of a vehicle was not, it seems, so closely tied to social bias and cultural perceptions of its usefulness to its owner.

Thus far, only individuals found in both the tax and probate records have been considered, but many individuals included in this 1863 sample group who did not own an enumerated pleasure carriage owned other types of vehicles. As would be expected, these men and women usually had estates of moderate or little worth compared to those containing the more expensive pleasure carriages. The $33,412.89 estate of North Hempstead's William Titus was an exception, although his inventory was not: three farm wagons ($15, $20 and $30), one light wagon ($40), a market wagon ($30) and four horses collectively valued at $435. For a few farmers with more modest estates the list was equally long. Charles Walter's estate of $1,932.50, for example, included three farm wagons ($80, $40 and $10), a large top wagon ($15), one large two-horse wagon

($75), a large spring market wagon ($100) and three horses collectively valued at $400. More typically, small farmers owned fewer vehicles, of which half or more were usually quite old and probably of little if any use. Timothy Whitemore of Hempstead, for instance, had an estate worth $1,495.60 that included two horses ($45 and $25) and four wagons with values of $50, $13, $10 and $6. Similarly, Benjamin Smith, also of Hempstead, with an estate worth $3,834.17, owned two horses ($60 and $50), a (possibly very old) sulky ($12), two farm wagons ($35 and $12) and a one-horse spring wagon valued at only $5. The poorest estates, as would be expected, contained few if any vehicles, and most of these were of little value. John Akley, also from Hempstead, had an estate listed at $106.50 that included a single top wagon worth only $10. Elbert Duryea's $417.17 estate included three wagons valued at $12, $10 and, remarkably, $2. Henry Combs, with property valued at $132.85, owned a pleasure sleigh worth $3 as well as a $10 spring wagon. Many others had no vehicles at all.

Undoubtedly, the old and inexpensive vehicles listed in the probate inventories represent the holdings of many Long Islanders who lived a lifetime on the farm, people who bought, built or refashioned usable vehicles while discarding old ones. Collectively, the variety found in these records illustrates the centrality of horse-drawn transportation at the core of agricultural life on the island. In addition to the vehicles, probate inventories listed wagon seats and sides, wood from broken farm wagons, broken or unused wheels, sleigh runners, discarded carriage tops and other artifacts of the age. Put simply, the tax records document ownership patterns among those most able to keep pace with the changing times, while the probate inventories provide a glimpse at many who collected the residue of a lifetime on the farm.

HARBINGER OF CHANGE: INDUSTRIALIZATION

The word *industrialization* conjures up familiar Dickensian images of de-skilled workers pitted against the omnipresent force of an increasingly mechanized, depersonal-ized society. Macrohistories of industrial America often provide similar images. Capitalists and capitalism, technological innovation and change, emerging networks of bureaucrats and professionals, standardization, the rising middle class–these are all common themes historians of late-nineteenth-century America turn to when examining the coming together of the nation after the Civil War. The development of Long Island's carriage industry during the third quarter of the nineteenth century, however, tells a different story. During this period the industry's development was indeed symptomatic of the larger region's socioeconomic transition from an almost exclusively agrarian orientation to one influenced increasingly by the forces of industrialization. But such truisms should not overshadow the circumstantiality of local conditions. We must qualify the all-encompassing impression that Northern entrepreneurs, to quote Daniel T. Rogers, "turned the land into a stupendous manufacturing workshop."[52]

To understand the impact of industrialization at the level of the artisan and his laborers we must look beyond technological determinism and its impact on economic life to examine the contextual dynamics that mediated change and gave to late-nineteenth-century developments identifying characteristics of both the old and the new, traditional practices and modern technique. In Suffolk County particularly, the shop economy that dominated the first half of the century remained strong. Industrial potential did provide new opportunities, but the resilience of long-standing practices, efforts to compromise with newer economic trends, the restrictions of a rural environment, economic necessity and the particular demands imposed upon the carriage making trades placed the island's more successful carriage manufacturers in a middle position, somewhere between the large-scale urban producer and the local blacksmith or wheelwright. Restricted by urban competition and limited largely to its own rural market, Long Island's carriage industry retained its rural characteristics, incorporating at once both industrial techniques and community service, comparatively large-scale production and personalized repairs. In this case, if

E. Tuthill & Co.,

CARRIAGE MAKERS,

PORT JEFFERSON, L. I.

Having again enlarged their factory and greatly increased their facilities for manufacturing Carriages, have on hand the largest and best assortment ever offered for sale in Suffolk or Queens counties, consisting of all made in the latest style, and of the best material.

Those in want of anything in this line are invited to call before purchasing elsewhere.

SECOND-HAND CARRIAGES

taken in exchange, and for sale.

E. T. & CO. keep on hand an assortment of light harness, single and double; also Horse Blankets, Fly-nets, Robes, Whips, &c., all of which are offered at moderate prices.

Painting, Trimming and Repairing done with Neatness and Despatch,

Port Jefferson L I., Nov. 13th 1866.   (n5yl)

ROCKAWAYS, PHAETONS, TOP AND OPEN BUGGIES, ONE, TWO & THREE SEAT TOP AND OPEN WAGONS, TROTTING WAGONS, SKELETONS, SULKIES, SLEIGHS, FARM WAGONS, &C,

BUY THE BEST.

Advertisement for
E. Tuthill and
Company,
Carriage Makers
June 21, 1887
The Long Island Star

there existed a "stupendous manufacturing workshop," it was manned largely by mechanics who retained ideals of skilled craftsmanship and personalized service. Within the island's most agricultural region, Suffolk County, the carriage industry kept pace with the changing times while retaining the basic values of the shop economy. Only in the last two decades of the century did dramatic changes begin to emerge.

Suffolk's carriage makers throughout the third quarter of the nineteenth century were a diverse lot. Local blacksmiths and wheelwrights, small-scale entrepreneurs and factory owners all contributed to the production of horse-drawn vehicles. Collectively, however, the first category–blacksmiths and wheelwrights–remained the dominant producers. A few manufactories, such as Tuthill and Company of Port Jefferson, would emerge as a community's industrial centerpiece, but they did not produce more than the local network of small shops. In fact, during its 20 years in operation–1855 to 1875–Tuthill's factory was a county-wide centerpiece as well. In 1870 the factory's 18 employees and $30,000 worth of products produced were approximately three times the levels of his closest competitors. Accordingly, Tuthill and Company alone was listed as a manufactory in that year's federal Products of Industry report.[53] In 1880, almost five years after the collapse of Tuthill and Company, no Suffolk manufacturer of horse-drawn vehicles could approach these numbers. In that year, Suffolk's five largest producers (two of whom are designated blacksmiths) collectively produced products valued at $39,500, only about one-third more than Tuthill's factory alone had produced one decade earlier. The single greatest contributor to this 1880 figure, Huntington's Wm. T. Downs, employed nine workers, exactly half Tuthill's 1870 work force, and had a production level of $10,000, only one-third Tuthill's mark.[54]

These figures indicate clearly that Suffolk's carriage manufacturers remained in a modest position. They were able to produce on a level that distinguished them from the independently employed craftsman who usually hired no more than one or two workers, but they were unable to compete on a level anywhere equivalent to their urban counterparts. As mentioned earlier, some reasons for this are obvious. Still, why is it that no manufacturer emerged to dominate Suffolk's carriage manufacturing industry despite the longevity of many producers? Why did Tuthill's 1870 levels remain an anomaly throughout the period? For answers we must turn to those conditions specific to the rural environment.

First, the shop economy that dominated the carriage making trades on the island influenced the competitive posture assumed by the larger manufacturers. The informality of the small shop, its promise of personalized service, a mechanic's reputation for craftsmanship–these were all traditional inducements offered to the consumer by the local shop. To exist in the same market, larger manufacturers could not forego these competitive advantages. Throughout the late nineteenth century their advertisements constantly assured the consumer that personal service, attention to detail, and the highest level of craftsmanship were among their foremost considerations. Although some producers would actively compete with the high-volume urban manufacturing image, their financial stability depended upon community service; they still relied on the shop economy image of dependability and personalized attention. Tuthill, for example, could boast of having the "largest and best assortment of Carriages ever offered for sale at any establishment in Suffolk or Queens Counties," yet he still promised prospective buyers that all work would be done under his own close supervision. "The best class of work" and "none but first class materials" were assured; "any style that may be desired" was available upon request; and services including painting, trimming and repairing would be "done with neatness and dispatch."[55]

This last area, listed in the census data as "repairs," "jobbing," or simply "other work," contributed significantly to the manufacturer's profits and served as a fundamental link to the material base of the shop economy. Most carriage manufacturers, beginning their careers as blacksmiths or wheelwrights, simply expanded their shops to include facilities and tools necessary for increased production of horse-drawn vehicles.[56] This developmental process retained a traditional dependence on nonmanufacturing work to supplement their income, to support a successful expansion and to help them survive economic hard times, a frequent occurrence during the second half of the nineteenth century.[57] As evidenced by the 1860 Products of Industry report, repairs and jobbing accounted for more than 55% of all work done by Suffolk County producers of horse-drawn vehicles whose products are itemized in the data. And there is only a minor difference between the large and small producer, the four largest averaging almost 47%. After a decade of industrial growth, this percentage dropped to 32%. Of Tuthill and Company's $30,000 total, 22% accumulated from repairs. Later census data do not itemize these figures, but the economic disasters of the mid-1870s, among other factors, limited

Port Jefferson, Long Island, New York
late nineteenth or early twentieth century Photographer, A. F. Davis (?) Photograph Courtesy of The Long Island Collection, The Queens Borough Public Library, Jamaica, New York

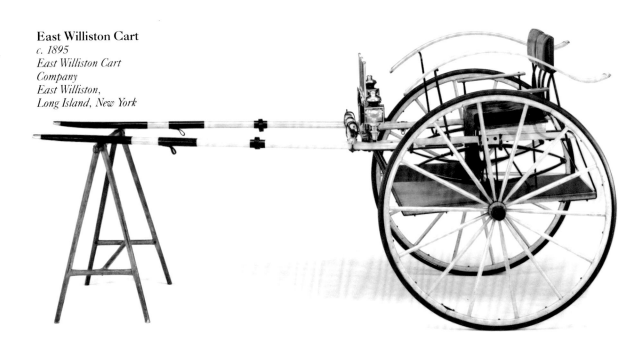

East Williston Cart
c. 1895
East Williston Cart
Company
East Williston,
Long Island, New York

additional growth.[58] In effect, the 1870 levels represent a stabilized ratio between carriage production and nonmanufacturing work, at least until reputations based on specific carriage designs were established during the next decade. Throughout the remainder of the nineteenth century, those carriage manufacturers who advertised in local papers continued to list among their services an ability to accommodate all forms of nonmanufacturing needs.

One problem that faced all of the island's carriage producers was the quickly saturated local market. By the 1870s most Long Islanders who could afford a carriage had one, and the market's ability to consume increased production was indeed limited. The observer from Huntington who noted in 1875 that the "spirit of Progress" brought with it a spirit of competition for the latest and most fashionable carriages gave convincing testimony to the growing proliferation of vehicles on the island. But with a saturated market–and lacking a national or regional market–no manufacturer could hope for the massive expansion so characteristic of industrialized America. In fact, as the developments in Huntington illustrate, the increasing demand for the most fashionable vehicles gave a competitive advantage to the more prestigious cosmopolitan carriage designers. Once again, Tuthill and others were cast in an in-between position. If a fashion-conscious customer desired the latest cosmopolitan style, and cost was of secondary consideration, he or she might turn to Brewster & Company or one of the many urban manufacturers who advertised in local papers. If function and cost were primary concerns, a mass-produced vehicle assembled by a company such as Studebaker or an inexpensive one made by a local craftsman might be the logical choice. For local carriage manufacturers, a balance between these two became a working market strategy. One Huntington manufacturer, for example, advertised he would have "Carriages of every description made to order which for beauty and durability will compete with city work. Prompt attention paid to Painting, Trimming, and Repairing."[59] Perhaps more concerned with the competitive advantages of mass production, William T. Downs, who was establishing a growing reputation for fine craftsmanship by the late 1870s, boasted that "any type work [can] be turned out at City prices."[60]

Long Island manufacturers could not either customize or mass-produce vehicles competitively, so they offered a compromise: individualized attention and, to use Tuthill's words, "moderate prices."[61] As evidenced by their many advertisements in local tabloids, customization took the combined form of personalized service and production flexibility. In Tuthill's advertisement of May 17, 1873, for example, he provided a detailed list of available new and secondhand vehicles and advised that "Persons in want of anything like the above WILL DO WELL TO CALL AND EXAMINE THE STOCK. If you don't see what you want in the show room you can have JUST SUCH A ONE AS YOU LIKE BUILT ON SHORT NOTICE."[62] Coles & Bentley, also of Port Jefferson, were more succinct, stating simply that "CARRIAGES OF ALL DESCRIPTIONS [WILL BE] BUILT TO ORDER."[63] In a way, one could argue that a mechanic's reputation for craftsmanship was a form of compensation. His skills and personal attention would assure the finest work and accommodate all demands "at reasonable prices;" and, of course, he was available upon request. In effect, for the local carriage maker, a traditional artisanal orientation was a significant asset. In rural areas, as long as face-to-face transactions remained the dominant form of doing business, a durable shop economy could accommodate change without abandoning those values that still served it well.

By the 1880s and 1890s there were many developments that signaled a departure from the artisanal way of life. Of these perhaps the two most significant were the growing importance of inventive designs and the increasing use of prefabricated or interchangeable parts. After the economic recession of the 1870s had passed, several Long Island manufacturers were able to rise to prominence in the market by designing their own patented carriages; the most successful of these were the various Long Island road carts.[64] The mineola cart, first patented by Charles A. Ellison of Mineola in 1885, was a convenient two-wheel pleasure vehicle that quickly became very popular. Similarly, the east williston cart, patented by Henry S. Willis in 1891, and the hempstead cart, manufactured in 1900 by Robert H. Nostrand, contributed to the growing reputation of Long Island carriage manufacturing. With these popular designs, carriage establishments could expand to accommodate the new demand for their products. None, however, was more active in this regard than Riverhead's Charles Blydenburgh. From 1880 to 1890 he patented 23 items: six vehicles, eight vehicle springs, one wagon

hound, two carriage poles, three vehicle gears, two shifting seats and a vehicle body.[65] His most successful designs by far were the montauk and suffolk wagons, vehicles that brought Blydenburgh a small reputation outside the local marketplace. As reported in the March 1886 issue of *The Carriage Monthly*, demand for his vehicles

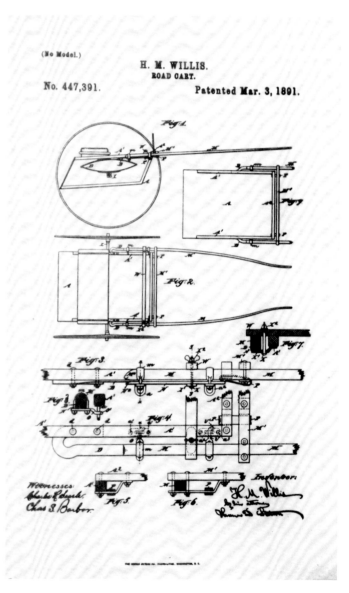

Road Cart,
Patent No. 447391,
Patented March 3, 1891
Developed by Henry M.
Willis of East Williston,
New York
Copy Courtesy of
U. S. Patent Office

was rather high: "C. M. Blydenburgh, Riverhead, Long Island, New York, reports running full force of men winter and summer since the introduction of his specialty, the 'Montauk' wagon, for which he finds a steady sale in many parts of the United States."[66] By this date, Blydenburgh was eastern Long Island's most prodigious producer.

Others, such as E. C. Lefferts, could profit by a single invention. His wagon spring, it was reported, enabled "a farmer to turn his old-fashioned farm wagon into a spring wagon at very short notice." Obviously, here was a sound investment:

It would pay for itself in use to a farmer in one season for the one purpose alone of carting potatoes and apples in barrels. They are usually shaken down so much on a trip to the packets that a large number of extra ones have to be taken along to supply the deficiency. But the comfort of riding in a spring wagon when a farmer has been shaken over stony roads nearly all his life in an old-fashioned wagon is too evident for comment.[67]

Lefferts profited considerably from his wagon spring and eventually sold the rights of distribution to Smithtown's A. E. Hallock. His invention was now available wherever Hallock advertised his wares. This increasing use of innovative designs as well as specialized springs, seats, gears, wagon bodies and related items was in turn part of the second development mentioned above, the carriage maker's growing use of ready-made parts. No longer were craftsmen manufacturing their products exclusively from raw materials. According to conventional wisdom, this is proof positive that a new age had arrived, and, in a limited sense, it had. Rather than stressing how much this development distanced the artisan from tradition, however, we can note how the use of prefabricated parts was incorporated into traditional practices. In this case, despite the altered manufacturing process, the mechanic-oriented system of production remained. Using prefabricated parts was simply another means of competing with fashionable designers without abandoning the values of a shop economy. If anything, this development strengthened the resilience of long-standing practices by providing a mechanism for successful competition that

The Blydenburgh "Montauk" Wagon, Improved
c. 1900
From catalog, Specialties
C. M. Blydenburgh
Riverhead, Long Island,
New York

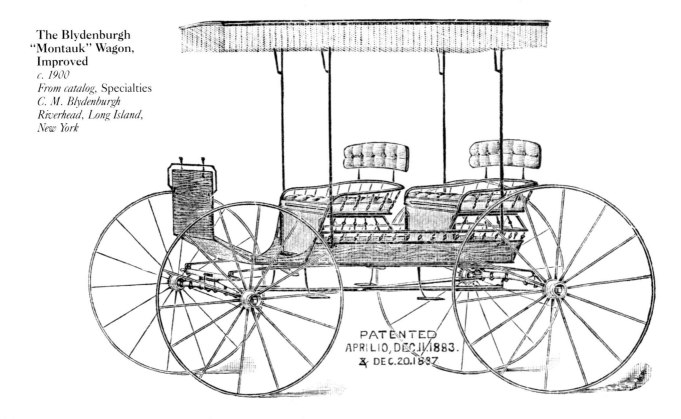

PATENTED
APRIL IO, DEC.II,1883.
& DEC.20,1887

did not reduce the social significance of artis-anal trade. Without compromising their reputations, skilled mechanics could pool their talents and use parts manufactured elsewhere to produce vehicles in a fashion quite removed from traditional manufacturing practices, as the following description shows:

Wm. Sammis has a very tasty little sulky, which does credit to the builders, as this sulky is the one of the kind ever built in this place [Babylon], it deserves more than a passing notice. The woodwork was done by S. Hilton, the famous carriage builder of Albany, and is strongly, and at the same time lightly built. The iron work was done by Sammis & Cornelius, and the painting by A. I. Moore. The running part is blue, with bright yellow stripes and black trimming, with the hubs, and tips of the shafts, as well as the railing around the cain seat, are silver plated. The entire sulky weighs 75 pounds and compares favorably with any we have ever seen.[68]

In this case, it would seem, local reputation was as much a selling point as the finished product. Although it might take a combined effort and specialized skills to remain competitive with an urban manufacturer like Brewster & Company, that effort remained characteristic of a rural production system that retained its own unique dynamic.

Another cooperative enterprise of a different type, partnerships, was also quite common in Suffolk County during the third quarter of the nineteenth century. Rarely, however, did these remain intact for more than a short period of time. Cooperatives were also a form of compromise, a means of combating the transformative dangers associated with industrialization. As Daniel T. Rogers has shown, reform literature told of the reductive potential of modern economic forces, the inevitable de-skilling of the handicraft trades.[69] Facing this threat, skilled mechanics, if they combined their abilities, could effectively retain their status and trade. Using Rogers's words, "cooperation promised independence."[70] Like the department store–a contemporaneous development–the cooperative promised large-scale growth through addition, not transformation. On Long Island, wheelwrights joined with blacksmiths, wagon makers with carriage painters, carriage smiths with trimmers and

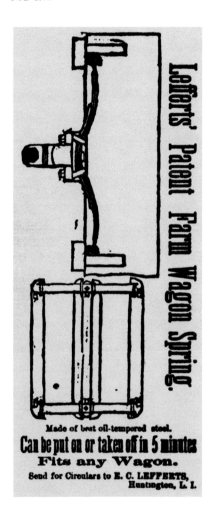

harness makers, and so on. Most soon found out that profits did not accumulate accordingly; having to split profits rather than pay wages, they learned, was far too costly. Despite its ephemeral nature, however, in the cooperative we see once again the artisan's allegiance to his trade.

As mentioned above, specialization also became an identifying characteristic of the carriage making trades during this period,

Bersenger Carriage Builder
c. 1890
Huntington,
Long Island, New York
Photograph Courtesy of
The Long Island Collection,
The Queens Borough
Public Library,
Jamaica, New York

especially in the factories that began dominating carriage production during the 1880s and 1890s. Earlier, Tuthill had been alone in this and had failed, but after the economic hard times of the 1870s several carriage manufacturers expanded their operations and built factories capable of large-scale production. In these the division of labor characteristic of factory-style production was in every way apparent. A. E. Hallock's new factory, built in 1881, was typical: "Mr. Hallock's factory has three floors and all branches of the work are carried on in the establishment, there being separate departments for blacksmith, wheelwright, painting and varnishing, and trimming."[71] Although this division would retain in part the artisanal orientation, and the proprietor could boast that he kept "on hand a complete force of mechanics" independently working in the various departments, these craftsmen were themselves now part of the factory machine.[72] As one observer noted, despite Hallock's departmental approach, "everything is done by *system* and . . . by means of an ingenious hoistway the carriages and parts of carriages are raised and lowered from one floor to another with great rapidity and ease."[73] In effect, the mechanic-oriented system of production was now housed in a single building where efficiency became the organizing principle. The mechanic's reputation for craftsmanship, though, retained its role in the marketplace. In 1885, the *Long Islander* described in considerable detail how the incorporation of the mechanic in the factory system provided Babylon's A. A. Cornelius with the necessary competitive edge. The "horse-shoeing department" was run by Eli Anderson and Mickey McCarthy, the latter being "a horseshoer of note, having charge of the shoeing of Mr. August Belmont's noted running horses for years." And, it was reported, "Mickey's song and the sound of his hammer on the anvil ring out merrily each day from morn till night. More first-class horses are shod here than in any other shop in Suffolk County." Other departments were equally well manned:

The carriage smithwork is in the charge of Andrew J. Weeks, an efficient and first class carriagesmith with his two assistants. John Hilton bosses the wood shop, and has competent men to help rush through work in that department. John Place has no superior as head of the trimming shop, and in trimming and upholstering fine coaches and carriages is kept busy. Frank Holdridge runs the painting of all work brought to the establishment independently, yet is situated in the same building. The very fine work that Frank and his assistants turn out, continually brings new patrons, while the old all appreciate their work coming from his shop. Mr. Cornelius oversees the whole business, and by his obliging business manners, good work and moderate charges he is enjoying the bulk of trade in his line in this vicinity.[74]

As indicated, carriage painting in particular became a highly valued, specialized occupation, combining the aesthetic sensibilities of the fine artist and, if independent of the factory, those skills necessary for successful operation of an artisanal shop. To gain access to the marketplace a painter would first have to establish a reputation for artistry. Often, as in the case of Timothy Scudder of Huntington, this was done in areas unrelated to the carriage industry–landscape or portrait painting, for example, might provide the necessary proof of ability. The emphasis, as implied by the many descriptive newspaper reports, was on deftness in capturing on canvas or carriage a replication of the real, the ability to "counterfeit nature."[75] Apprenticeship experience therefore was not, as in so many other trades, the predetermined first stage to success. Because carriage painting required the unique skill of the artist, learned skills could not compensate for the absence of artistic ability. An apprentice might gain essential experience, but that experience would not, in the eyes of the public, qualify him for success. The ability to paint "in the highest style of the art" was alone the most valued skill, regardless of how it was developed. In what could almost be described as a public-relations campaign, the *Long Islander* actively promoted the artistry of young Scudder, who was also, not coincidentally, its first advertiser of a carriage-related trade after more than a year of economic depression. We can reasonably assume that Scudder's claim of experience and his willingness to guarantee customer satisfaction were secondary assurances, not primary inducements. For the latter he let his artwork, or perhaps more accurately the newspaper reports that described it, provide proof of his ability.[76]

Because reputation was the primary asset of the carriage painter, local newspapers–always more than willing to promote the local economy and the particular skills of local craftsmen–usually mentioned the name of the artist when reporting on the assembly or repair of important or unusual vehicles.

Grocer's Advertising Wagon
late nineteenth or early twentieth century
Patchogue,
Long Island, New York
Photograph Courtesy of The Long Island Collection, The Queens Borough Public Library, Jamaica, New York

For example, when reporting on the repair of a butcher's wagon at A. E. Hallock's Smithtown shop, the *Long Island Leader* noted that "Charles Marvin done the painting, and C. Sanford the lettering. The wagon presents a handsome appearance and reflects great credit upon the workmen."[77] Similarly, I. M. Moore, who painted the carriage for Wm. Sammis cited above, was frequently mentioned in descriptive accounts that detailed the high-quality work done by local carriage makers. In fact, for a vehicle to have been found worthy of specific mention in a newspaper account it must necessarily have displayed the distinctive mark of artistic design.

By the 1880s, when many carriages and carriage parts were increasingly being shipped in from areas outside Long Island, carriage painting took on added significance as a distinguishing stamp of localized production. Prior to the third quarter of the nineteenth century a vehicle's worth was primarily determined by a craftsman's reputation for producing from scratch a finished product. Thus the initial stages of production–the process by which a vehicle's durability and structure were assured by the hands of the workman–were more valued. By the 1880s much of this work could be done, and often was done, by workers quite removed from the local market. Accordingly, the finishing of a vehicle–its artistry when openly displayed–assumed additional significance, particularly in a society increasingly preoccupied with cosmopolitan fashion and the display of personal wealth. At this time, the carriage painter had a distinct advantage, for he alone was able to provide the finishing touch, so to speak. Only he could give to an assemblage of prefabricated parts the stamp of uniqueness in an age of discriminating taste.

With fashion so intimately connected to the perceived worth of a vehicle, one can easily understand why carriage makers Hallock and Vail, in the summer and fall of 1885, advertised with a note of desperation that "A First Class Carriage Painter [is] wanted at once."[78] Much to the chagrin of Hallock and Vail and other local carriage manufacturers, skilled painters increasingly found the marketability of their skills such that independent enterprise was not only possible but highly profitable. As evidenced by the expanding career of Timothy Scudder, carriage painting offered opportunities not available at mid-century; his abilities as an artist and entrepreneur went hand in hand. In addition to painting vehicles for private customers, he free-lanced for several local carriage makers, working on the production and repair of locally manufactured vehicles; as an agent for more distant manufacturers he customized and sold carriages not of local design; he either built or assembled from prefabricated parts carriages in his own shop; he continued to profit from his reputation as a portraitist; and, when photography became a potentially profitable enterprise, Scudder was in a position to apply his talents, and increase his income, in a related and growing artistic trade.

Scudder's career illustrates the advantages specialization offered to those with ambition whose skills were independently valued as unique or God-given. In the factory, on the other hand, where most skills could be learned, the division of labor did not offer a mechanic similar opportunities for economic independence. Here, he applied his expertise within a system that competed with the shop economy while at the same time appropriating its values. The mechanic, in short, was working efficiently for a system of production that was gradually displacing the artisan shop. Still, the many skills incorporated in the making of a carriage strengthened the industry's ability to maintain mechanic-oriented production. In addition, as long as face-to-face transactions remained the norm in many communities, carriage makers would work and live on a scale comparatively equal to that of the clientele each of them served, a condition not enjoyed by those in the majority of other trades. As such, carriage makers were not plunged into the depersonalized world of industrialization that inhabits many antiquated textbooks. The manufacturer with the best reputation would obtain work on more expensive carriages from a higher-class clientele; he would, in short, rise with the price of his product. The inroads of industrialization and mass production would eventually establish profit criteria based on the number of items sold, and thereafter profitability would no longer be determined by the individual sale. At this stage, a manufacturer would no longer be dependent upon the rich for his riches and the social significance of individualized transactions would be systematically reduced; the many factories that began to prosper in the 1880s were indeed harbingers of change. But for many Long Island carriage manufacturers, this development lay in the future; their economic concerns remained wedged between the rural requirements of a still agrarian region and the demands of very different times.

The "spirit of progress" so evident in the changing patterns of carriage production, ownership and use of horse-drawn transportation and Long Island society generally were all part of the complex process of development that defined nineteenth-century life on the island. Vehicles found on a village farm, at the railroad station, in a livery stable or outside a church, on the many country roads or on the lot of a carriage manufacturer, reflected in a real way the *movement* of local society. As the new century

approached, factories were replacing artisans' shops as the dominant producers of horse-drawn vehicles, new and more elaborate vehicles were found with increasing frequency all across the island and the insular communities of the early nineteenth century had become part of the more commercial culture of the region. During the first years of the twentieth century, however, many of these dramatic changes would themselves blend into the twilight of a passing age, and Long Island streets would be increasingly shared with, and then dominated by, the inanimate objects of automation.

Lake Ronkonkoma
1900
Photographer, Henrys (?)
Lake Ronkonkoma,
Long Island, New York
Photograph Courtesy of
The Long Island
Collection,
The Queens Borough
Public Library,
Jamaica, New York

APPENDIX

TABLE 1:
Listing and Quantity of All Horse-Drawn Vehicles by Specific Description

*Totals for all entries that are not vehicles are in parentheses and are not included in the summary totals given for each period.

	1815–16	1837–38	1862–63	1881–82	1897–98
Horse	(157)	(161)	(140)	(96)	(88)
Chair carriage	1	–	–	–	–
Carriage	2	1	11	5	8
Stage	2	2	–	–	–
Wagon	81	72	29	18	5
Farm wagon	1	29	66	31	22
Phaeton	1	–	–	3	4
Sleigh	11	10	18	12	9
Pleasure sleigh	10	14	16	5	1
Box sleigh	3	–	–	1	–
Sled	11	6	3	2	2
Pleasure sled	2	1	1	–	–
Wood sled	26	24	7	4	5
Box sled	3	1	1	–	–
Stone sled	1	–	–	–	–
Wood sleigh	7	3	10	3	2
Cart/horse cart	4	4	7	12	12
Wagon sides (pair)	(10)	(5)	(9)	–	(1)
Wagon seats	(5)	(4)	(4)	(1)	–
Riding chair	(16)	(5)	–	–	–
	166	167	169	96	70
Mule	–	(2)	(2)	(2)	–
Closed carriage	–	2	1	2	–
Extension top carriage	–	2	–	–	–
Barouche	–	2	3	–	1
Sulky	–	3	3	1	1
Gig	–	4	–	–	–
Buggy (or buggy wagon)	–	2	9	2	4
2-horse farm wagon	–	2	3	4	–
Large wagon	–	5	8	1	–
Pleasure wagon	–	14	1	–	–
1-horse wagon	–	8	10	2	2
2-horse wagon	–	7	4	–	3
Covered wagon	–	1	–	–	–
Top wagon	–	6	16	4	3
1-horse sleigh	–	2	1	–	1
Top sleigh	–	1	–	–	–
Slab sled	–	2	–	–	–
Hay cart	–	1	–	–	–
		231	228	112	85
Pony	–	–	(2)	–	(1)
City coach	–	–	2	–	–
Light coach	–	–	1	–	–
Heavy coach	–	–	1	–	–
Omnibus	–	–	1	–	–
4-passenger carryall	–	–	4	–	–
Top carriage	–	–	1	1	–
2-horse carriage	–	–	5	–	–
1-seat phaeton	–	–	1	–	–
Canopy	–	–	1	–	–
4-seat phaeton	–	–	1	–	–
Rockaway (or rockaway wagon)	–	–	6	6	3
1-seat rockaway	–	–	2	–	–
2-seat rockaway	–	–	1	1	2
2 seat top rockaway	–	–	1	–	–
3-seat rockaway	–	–	1	–	–
Open buggy	–	–	4	1	1
Top buggy	–	–	10	15	6
2-horse spring market wagon	–	–	4	–	–
Top market wagon	–	–	2	–	–
Express wagon	–	–	1	1	–
Business wagon	–	–	1	3	9
Truck wagon	–	–	1	–	1
Iron ale farm wagon	–	–	3	–	3
Potato wagon	–	–	1	–	1
2-horse potato wagon	–	–	1	–	–
2-horse hay wagon	–	–	1	–	–
Manure wagon	–	–	1	–	–
2-horse manure wagon	–	–	1	–	–
Feed wagon	–	–	1	–	–
Light 2-seat wagon	–	–	1	–	1
Calash top wagon	–	–	4	–	–
2-horse top wagon	–	–	1	–	–
1-seat top wagon	–	–	1	–	–
Driving (or riding) wagon	–	–	2	–	–
1-horse platform spring wagon	–	–	1	–	–
1-horse spring wagon	–	–	17	–	2
Box wagon	–	–	3	6	4
Box spring wagon	–	–	7	–	–
Grocery wagon	–	–	3	–	–
Meat wagon with top	–	–	2	–	–
Market wagon	–	–	11	8	9
Stage market wagon	–	–	1	–	–
Spring market wagon	–	–	4	–	–
2-horse sleigh	–	–	3	–	–

	1815–16	1837–38	1862–63	1881–82	1897–98
Large sleigh	–	–	2	2	–
Stone sleigh	–	–	1	–	–
Business sleigh	–	–	1	1	–
Farm sleigh	–	–	1	1	–
Side saddles	–	–	(3)	–	–
			355	158	127
Carryall	–	–	–	2	–
2-seat phaeton	–	–	–	1	–
Park phaeton	–	–	–	1	–
Charlotte	–	–	–	1	–
Heavy carriage	–	–	–	1	–
2-seat carriage	–	–	–	2	3
Coach	–	–	–	1	–
1-horse farm wagon	–	–	–	4	3
3-horse farm wagon	–	–	–	1	–
Iron axle wagon	–	–	–	1	–
Hay wagon	–	–	–	1	6
Skeleton wagon	–	–	–	1	1
2-seat wagon	–	–	–	4	–
2-seat open wagon	–	–	–	1	1
Open wagon	–	–	–	2	1
2-horse open wagon	–	–	–	2	–
3-seat top wagon	–	–	–	1	–
Road wagon	–	–	–	4	–
Spring wagon	–	–	–	5	4
Platform spring wagon	–	–	–	1	1
2-seat spring wagon	–	–	–	1	1
1-horse spring box wagon	–	–	–	2	–
1-horse box wagon	–	–	–	1	–
2-horse box wagon	–	–	–	1	–
Top grocery wagon	–	–	–	1	–
Milk wagon	–	–	–	2	1
Shaving wagon	–	–	–	1	–
1-horse lumber wagon	–	–	(3)	4	–
Wood wagon	–	–	–	1	–
Coal wagon	–	–	–	1	–
2-seat sleigh	–	–	–	2	2
Heavy sleigh	–	–	–	1	–
Bob sleigh	–	–	–	2	–
Farm Sled	–	–	–	3	–
Double runner sled	–	–	–	1	–
Dump cart	–	–	–	2	6
				221	157
Surrey	–	–	–	–	2
2-seat surrey	–	–	–	–	2
Extension top surrey	–	–	–	–	1
Trap	–	–	–	–	1
Run about	–	–	–	–	2
2-seat trap wagon	–	–	–	–	1
Summer carriage	–	–	–	–	1
Top phaeton	–	–	–	–	1
Extension top phaeton	–	–	–	–	1
Pony phaeton	–	–	–	–	1
Coupé rockaway	–	–	–	–	1
Calash top buggy wagon	–	–	–	–	1
1-horse buggy	–	–	–	–	2
					174
Sidebar buggy	–	–	–	–	2
Side spring buggy	–	–	–	–	1
3-seat box wagon	–	–	–	–	1
Buckboard wagon	–	–	–	–	3
Spindle wagon	–	–	–	–	1
Light top wagon	–	–	–	–	2
2-seat top wagon	–	–	–	–	1
1-horse top wagon	–	–	–	–	1
1-seat road wagon	–	–	–	–	2
Sidebar wagon	–	–	–	–	1
Broad tire farm wagon	–	–	–	–	1
Open business wagon	–	–	–	–	1
Double truck	–	–	–	–	2
Meadow wagonf	–	–	–	–	1
Mill wagon	–	–	–	–	1
2-horse spring wagon	–	–	–	–	1
High seat spring wagon	–	–	–	–	1
2-spring wagon	–	–	–	–	2
Depot wagon	–	–	–	–	2
Top depot wagon	–	–	–	–	2
2-seat sidebar depot wagon	–	–	–	–	1
Cuttery sleigh	–	–	–	–	1
Double runner sleigh	–	–	–	–	1
Bob sled	–	–	–	–	2
Road cart	–	–	–	–	5
Mineola cart	–	–	–	–	2
Breaking cart	–	–	–	–	1
Low cart	–	–	–	–	1
					217

NOTES

1 For information concerning Long Island's history and development, see Benjamin F. Thompson, *History of Long Island: From Its Discovery and Settlement to the Present Time* (1918; reprinted Port Washington, New York: Ira J. Friedman, Inc., 1962); Nathaniel S. Prime, *A History of Long Island: From Its First Settlement By Europeans to the Year 1845* (New York: Robert Carter, 1845); Peter Ross, *A History of Long Island: From Its Earliest Settlement to the Present Time* (New York: Lewis Publishing Co., 1902); Ralph Henry Gabriel, *The Evolution of Long Island: A Story of Land and Sea* (New Haven: Yale University Press, 1921); Jacqueline Overton, *Long Island's Story* (Garden City, New York: Doubleday, Doran & Co., 1929); Bernie Bookbinder, *Long Island People and Places: Past and Present* (New York: Harry N. Abrams, 1983).

2 C. H. Schmidt, *How to Develop Long Island* (Riverhead, New York: Roanoke Press, 1890). The only known copy of this work is located at The Richard H. Handley Long Island History Room, The Smithtown Library, Smithtown, New York.

3 Ross 280.

4 Gabriel 146.

5 Brookhaven Town Records, Brookhaven Town Clerk Offices, Patchogue, New York, envelope 13, no.1.

6 Brookhaven Town Records, envelope 13, no.2.

7 Gabriel 146.

8 Ross 281.

9 Gabriel 149.

10 Ross 281.

11 Prime 54.

12 L. Bailey, Letter to J.[?] Lawrence Smith, 8 September 1857, ms., The Richard H. Handley Long Island History Room, The Smithtown Library, Smithtown, New York.

13 Reprinted in Gabriel 151-2.

14 *Long Islander* 12 March 1875.

15 *Long Islander* 25 October 1873.

16 *Long Islander* 4 March 1881.

17 Federal census Products of Industry Reports for 1850 and 1860 and the Reports on Manufacturing for the years 1810, 1820 and 1840 demonstrate how very limited industrial activity was on Long Island in all areas during the first half of the nineteenth century. Microfilm Collection, State University of New York at Stony Brook.

18 *Corrector* 9 November 1822; *Long Islander* 22 April 1842, 21 June 1845, 1 July 1842; *Queens County Sentinel* 29 July 1858.

19 *Long Island Star* 22 January 1869; *Long Islander* 30 July 1887.

20 In 1841 the Long Island Railroad extended only to central Suffolk, ending at the "Suffolk Station," then located between Islip and Smithtown. By the 1870s several additional lines had been completed that connected almost every area on the island to New York City.

21 *Long Islander* 16 January 1885.

22 *Long Islander* 13 February 1885.

23 Reprinted in the *Long Islander* 15 August 1875.

24 *Corrector* 5 January 1827, 3 March 1844.

25 *Long Islander* 1 January 1858, 14 May 1858; *Queens County Sentinel* 7 January 1864; *Long Island Leader* 27 March 1875.

26 *Long Island Leader* 21 August 1875.

27 *Long Island Leader* 22 April 1882, 10 October 1884, 17 June 1881, 19 September 1884.

28 *Queens County Sentinel* 19 June 1884.

29 *Long Islander* 30 April 1875 (emphasis added).

30 Reprinted in *Long Islander* 8 June 1883.

31 *Long Islander* 29 May 1885.

32 *Queens County Sentinel* 22 May 1984.

33 *Queens County Sentinel* 28 April 1881.

34 *Long Islander* 5 August 1881.

35 *Picturesque Bay Shore, Babylon and Islip* (New York: Mercantile Illustrating Company, 1894) 19, 36, 67.

36 *Long Islander* 1 February 1867.

37 *Long Islander* 13 February 1874.

38 *Long Islander* 27 February 1874.

39 *Queens County Sentinel* 9 February 1882.

40 *Long Islander* 22 December 1876.

41 *Long Islander* 6 February 1880.

42 Abstracts of Valuations, for the New York State tax on real and personal property, 1800, New York State Archives, Albany. The author would like to thank Doris Halowitch for her painstaking tabulation of this data. The totals cited in the analysis that follows here are the products of her careful and meticulous research. See Appendix A in Halowitch's essay, page 108, for a description of this resource.

43 These percentages, of course, do not reflect family ownership patterns. Rather, they reflect ownership for the aggregate population, including every man, woman and child listed in the census data.

44 Assessment Lists of the Federal Bureau of Internal Revenue (1862), Rolls 38, 44, 49. Microfilm Series M 603, National Archives, Washington, D.C.

45 Federal population figures for the year 1860 are used here for the analysis below.

46 According to the New York State census of 1855, Sag Harbor was the only Suffolk village with a population over 2,500.

47 The tax records did not break down Smithtown Township or East Hampton Township totals, so village comparison for these areas was not possible.

48 *Queens County Court House Inventory of Estate* records. Microfilm, Eisenhower Park Museum, Nassau County.

49 The totals, of course, represent the total number of vehicles listed for each group of 100 inventories considered below. The manageable figure of 100 was used so that ownership could easily be expressed in percentages. By way of example, the total of 166 vehicles found in the 100 inventories considered for the years 1815-16 averages out to 1.7 vehicles per individual.

50 Table 1 is arranged specifically to illustrate this point.

51 *Long Islander* 28 November 1884.

52 Daniel T. Rogers, *The Work Ethic in Industrial America, 1850-1920* (Chicago: University of Chicago Press, 1978) xii.

53 Ninth Census of the United States (1870), Products of Industry, New York State.

54 Tenth Census of the United States (1880), Products of Industry, New York State. The other four manufacturers were Chas. Hammond (blacksmith, Babylon: 7 employees, $10,000); Chas. Blydenburgh (carriage maker, Riverhead: 6 employees, $7,500); Alanson E. Hallock (blacksmith, Smithtown: 5 employees, $6,000); and Coles, Baldwin, and Bentley (carriage maker, Port Jefferson: 8 employees, $6,000). These data suggest that Suffolk's more successful carriage makers, each working in separate towns, gained their status by developing superior reputations in their respective villages. Therefore, it would be difficult to expand their market base beyond their own communities, where only detached customer relations were possible.

55 *Long Island Star* 13 September 1867; *Independent Press* 17 June 1869.

56 A quick survey of the Products of Industry reports reveals that no distinct category "carriage maker" existed. However, the lure of carriage making as the most prestigious occupation among the related trades is evident. In the 1860 and 1870 reports there is little difference in terms of manufactured product among many who are listed as blacksmiths, wagon makers, wheelwrights and carriage makers. In several cases, those listing themselves as carriage makers identified themselves (or were identified by the recorder) not by the work they produced but by the work they hoped to produce. Curiously, many listed as carriage makers did not have a finished carriage among their finished products.

57 The 5 May 1876 edition of the *Long Islander* contains a description of business at the carriage shop of Wm. T. Downs that illustrates this last point. Despite the reporter's required optimism, it is clear that repair work, and not manufacturing, sustained the business through the existing economic hard times. "It is a cheerful sight to go into carriage shops of our village during these times of general complaint regarding dullness of trade, and see the amount of business now being done in them. Wm. T. Downs has at least 25 or 30 wagons of every variety now on hand for repairs, besides orders for new work." The month was May, and the agrarian economy moved along with the seasons. Despite economic conditions, as long as farmers had to prepare their wagons for the summer months carriage makers could ply their trade.

58 Intense competition also limited growth potential. A quick survey of the *Long Island News Letter* for the years 1884–86 identified as many as six New York City manufacturers who advertised their products alongside local producers in a single issue. In addition, auctions of used vehicles, carriage rentals and agents for out-of-state manufacturers all supplied local demand.

59 *Long Islander* 6 November 1868.

60 *Long Islander* 9 March 1877.

61 *Long Island Star* 13 September 1867.

62 *Long Island Leader* 17 May 1873.

63 *Long Island Newsletter* 30 August 1884.

64 For a detailed discussion of these vehicles see Tom Ryder, "The Long Island Carts, or The Mystery of the Meadowbrook," *Carriage Journal* 22.1 (Summer 1984):12-16.

65 Copies of Blydenburgh's most interesting patent specification documents and those of other Long Island carriage makers are located at The Museums at Stony Brook.

66 *Carriage Monthly* 21.6 (March 1886):347.

67 *Long Islander* 3 September 1880.

68 *Long Island Head-Light* 23 May 1874.

69 Rogers 28.

70 Rogers 41.

71 *Long Islander* 17 June 1881.

72 *Long Islander* 19 December 1884.

73 *Long Islander* 17 June 1881 (emphasis added).

74 *Long Islander* 31 July 1885.

75 For an example, see the anecdote entitled "The Power of Imagination," reprinted in the *Long Islander* 24 January 1868.

76 For example, see the *Long Islander* 31 December 1875; 9 March 1877; 6 April 1877; 1 February 1878; 30 August 1878; 6 December 1878.

77 *Long Island Leader* 4 September 1875.

78 *Long Islander* 26 June 1885.

BEFORE THE CART:

The Relationship Between Horses and Carriages

Merri McIntyre Ferrell

Of carriage horses . . . it might be said that their quality (if not their quantity) is an index of civilization; for the carriage horse changes his character from century to century, almost from year to year, as wealth and skill augment, as highways improve, as vehicles become lighter, as railroads are brought into play, as people use their steeds for pleasure and for show rather than for long and necessary journeys.[1]

One of the significant ways in which horses have been utilized is as an external source of power to pull wheeled vehicles; in this capacity, the horse has been used for probably close to 4,000 years, transporting both people and goods in vehicles that engage the horse's muscles and energy in an efficient and cooperative manner.[2] Although its role as a "draft," or pulling,

The Mail Coach
1878
From La Sellerie
Française et
Étrangere
Artist, Léné
Lithographer, A. Adam
Editor, Brice Thomas
Publisher, Louis DuPont
Paris, France

animal developed relatively late in the history of its domestication, the horse was well-adapted to this use. (The date of the domestication of horses varies among geographical regions. The earliest evidence found to date comes from a site in the Ukraine and is dated about 4000-3000 B.C., depending on the radiocarbon dating system used.)[3]

Literally to harness the energy of the horse, to engage its power in an effective way and to train it to obey commands required a thorough knowledge of the animal's basic physical and behavioral traits. Understanding the horse's psychology was important to the effectiveness of using him for a designated purpose: A driver needed to know what would cause a horse to balk or shy as well as what was the best technique to engage and direct its physical power, and was advised to understand not only the general psychology of the horse but also the personalities of individual

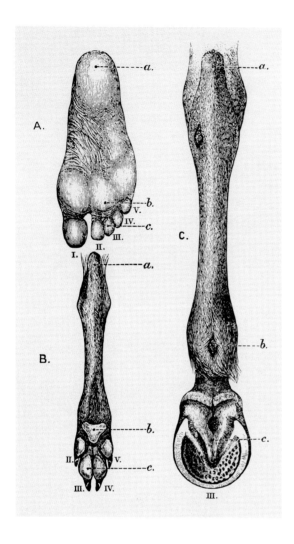

From **Modern Science:
The Horse**
1891
*William Henry Flower
Publisher, Kegan, Paul,
Trench, Trübner and
Company, Ltd.
London, England*

*These drawings illustrate the
plantar surface of the feet of
the following: A, human; B,
dog; C, horse. The numerals
I–V and letters a, b, c indicate corresponding points
on each foot.*

horses, each of which possessed its individual quirks or qualities. As recommended by
Price Collier, "Once you have a horse and
know something of his make-up inside and
out, and have housed him properly, and
bought his harness and learned something
of its use, the next thing is to make the
connection, first between the horse and the
vehicle, and then between yourself and the
horse."[4] Understanding the exertion needed
to pull a particular vehicle over various
surfaces and grades required a fundamental
knowledge of the potential and limitations
of a horse's physical characteristics and how
and why these characteristics developed over
time, as well as how the horse should be
maintained to maximize his efforts:

*A man on a box-seat with four reins in his hands,
who does not know how the horses in front of him
are housed, fed, shod, harnessed, and bitted, and*

*how by evolution they came to be what they are
physically and mentally, and the relative positions
of their vital organs and the bones of their skeletons; is not and will not be a coachman of any
competence until he knows something of these
things . . . . No man can bit a horse who knows
nothing of the inside of a horse's mouth; nor can
he fit him properly with his collar unless he knows
the relative position of the shoulder-blade and
humerus; nor can he see that his shoes are put on
to fit him unless he knows something of the formation of his foot; nor can he spare him fatigue and
help him through his simpler troubles on the road,
or in the stable, unless he knows something of the
horse's physical make-up.*[5]

To adapt physically to the environment
as that changed over millions of years, the
horse evolved from a small, multi-toed
animal to a large soliped with short power
muscles and long extensor limbs that
enabled it to cover a greater distance per
stride. The fact that the horse stands permanently on one toe–the hoof–also increases
its capacity for speed. As described in a
nineteenth-century text,

*The length and mobility of the neck, position of the
eye and ear, and great development of the organ
of smell, give them ample means of becoming
aware of the approach of enemies; while the length
of their limbs, the angles the different segments
form with each other, and especially the combination of firmness, stability, and lightness in the
reduction of all toes to a single one, upon which
the whole weight of the body and all the muscular
power are concentrated, give them speed and
endurance surpassing that of almost any other
animal.*[6]

The physical attributes of the horse, which were well adapted to the performance of tasks of power and speed, originally developed for survival purposes: The primary natural defense of the horse against predators was its ability to run swiftly.

The attributes of power and speed were also essential to the horse's services in association with man: The size, weight and proportions of the horse were well suited to its function as a riding or draft animal; its tractability facilitated both its initial domestication, and subsequent training to perform specific tasks. It has been suggested that, as a herd animal, the horse is inclined to submit to leadership. In its natural environment, allegiance is generally given to a dominant stallion; under the conditions of domestication, the tendency to submission is transferred to humans. According to Richard Smythe,

One of the most remarkable characteristics, common to nearly all horses, is a capability, even a willingness to transfer the allegiance and loyalty normally extended to another member of its species to a human being and to obey orders transmitted to it through various channels.[7]

The cooperative disposition of the horse was further modified through continuous association with human beings and through selective breeding that produced not only specialized physical characteristics and refinement of types, but also desirable behavior or temperament.

Because during the Carriage Era the horse's functional relationship to carriages was as a means of locomotion, it was regarded as a mechanical agent, and the discrete units of its specialized anatomy were analyzed as mechanical components.

An animal is but a beautiful piece of machinery, and although perfect in its construction, and wonderfully accommodating in its movements it still, like the engine, has a limited power, and has its peculiar modes of action, its strong and feeble parts; and we must well consider its structure, to

From Facts for Horse Owners
1894
D. Magner
The Magner Publishing Company, Battle Creek, Michigan

Some methods of training horses bordered on the absurd. Tripping a horse to force him into submission was recommended in many nineteenth-century books on horses. This method could result in serious injury to the horse.

How Would You
Break a Shier?
1892
From Gleason's Horse
Book
Oscar Gleason
Hubbard Publishing
Company, St. Louis,
Missouri, and Philadelphia
Pennsylvania

Oscar Gleason, who
demonstrated throughout the
United States his techniques
for breaking horses, offered
the following answer to the
question "How would you
break a shier?": "make him
thoroughly acquainted with
the beating of drums, the
rattling of tin pans, floating
the 'Star Spangled Banner,'
and the shooting of firearms,
fire crackers, music etc., by
driving him right up to them
and giving him to under-
stand he will not be hurt."

be able to apply the resistance in that degree, and
in that manner which shall enable it to produce
the greatest effect.[8]

The strength, action and structure of a
horse determined the extent to which it
could be utilized in harness. The rate of
speed and the distance a horse could travel,
its endurance and the weight it could pull
depended not only on the particular breed of
horse, but also on its general conformation,
fitness, soundness, age, diet and numerous
other influential factors. The performance of
a horse could be estimated, but not exactly
calculated. Although it was compared to a
machine or engine, it was still a living
creature, subject to fatigue, illness or injury.
It was capable of thought, memory and
emotions, if only to a limited degree.
According to H. C. Merwin, "There are
men to whom a horse is only an animate
machine: they will ride and drive him . . .
but all without a single thought of the
animal as having a character, a mind, a career
of his own, as being susceptible to pain or
pleasure."[9] Indeed, to those accustomed to

operating automobiles, it may be difficult to
comprehend that for hundreds of years the
motive power of land transportation was a
sentient being. The mentality and temperament of the horse were to be reckoned with
by those who enjoyed its services during the
horse-drawn Carriage Era. Establishing a
rapport with the horse was important, for

your horse is entitled to his opinion on the matter,
which you will do well to heed if you are to
continue amicable relations. Thus your preference
may be for a very leisurely exit from the stable
and a walk or very slow jog over the stones and
asphalt to the park. Your horse, however, is fresh
or nervous at some strange noises, or "jumpy," as
you sometimes feel yourself after a very long night,
too many cigars, and that last "highball" you did
not need. He wants to go on, and after the fashion
of a tyrannical man, you insist that he shall go
your pace. You pull, he pulls, and he makes a
discovery hitherto possibly unknown to him—that
you are not omnipotent and need not be implicitly
obeyed. This trouble of your own seeking results
not improbably in your discomfiture. It irritates
many a horse to be restrained for the first half-

mile or so and he is disagreeable all day if inter-
fered with, charming if humored.[10]

The analogy between horse and machine was not only a means of describing the function of its unique physical characteristics, but also provided a comparison by which the power of a horse could be measured. Observing the horse's anatomy and mobility contributed to its effective utilization as a motive power for vehicles and ultimately influenced vehicular and harness design as well as all other aspects of transportation, such as the surfacing and grading of roads. The relationship of the horse to carriages was extremely complex; the exact interaction between the muscular force of the horse and the mechanical structure of the vehicle was difficult to determine. Ideally, a carriage should not have any mechanical or structural elements that would impede the performance of the horse, nor should the harness confine the natural movements of a horse while in draft.

Draft referred to animal power applied as a mode of traction, the resistance of a particular vehicle and the surface over which animal and vehicle travelled: according to William Youatt,

there are three distinct agents and points of consideration in the operation of draught, which are quite independent of each other. They are—First, the moving power and the mode of applying it; secondly the vehicle for conveying the weight to be moved; thirdly, the canal, road, or railway, or what may be generally termed the channel of conveyance.[11]

Varying conditions affected the weight a horse could draw, the distance he could travel while pulling and the rate of speed he could maintain. "The distances which horses in perfect condition can go upon the road varies greatly with the powers of the animal, the degree of pains bestowed on him, the skill of the drivers, and the amount of his load, as well as the state of the roads."[12] The strength of individual horses also differed. To assess the power required to draw a particular vehicle would require measuring the strength of each individual animal employed in the task.

Definitions of draft appeared in literature published during the Carriage Era (c. 1700–c. 1920) to clarify, often in pseudo-scientific terms, the action of the horse as it drew a vehicle. In his "Treatise on Draught," Youatt expressed it as

the action of pulling [which] is effected in either case by throwing the body forward beyond the feet, which form the fulcrum, and allowing the weight of the body, in its tendency to descend, to act against the resistance applied horizontally, and drag it forward; as the resistance yields, the feet are carried forward; and the action renewed, or rather continued.[13]

William Bridges Adams identified the horse's muscular power and gravity as being the key components of draft:

When a horse is used for the purposes of draught, part of his power consists in the elastic action of his muscles, which serve to throw his body forward, as when he first bends and then straightens his limbs; and part of it in gravity, as when he hangs his body forward against the traces after the expansive muscular action is expended.[14]

One of the difficulties of defining draft, or horsepower, resulted in the fact that each person analyzing the subject selected different aspects of the various components affecting the action. Whereas Youatt described draft in terms of the horse's exertion and Adams defined it as being subject to muscular energy in conjunction with its resistance to gravity, J. Jacobs asserted that draft entailed the transmission of animal motion to a stationary vehicle "as the generation and continuation of local motion in a body at rest by another body, at first equally destitute of motion; the motion given to a body in traction not being communicated, as in percussion by another body already in motion."[15]

The term *horsepower* was first applied in 1783 by James Watt to describe the mechanical capabilities of his steam engine, after being asked by his customers "how many horses his engines would replace." Following experiments with powerful draft horses, he discovered that one horse "travelling at 2 1/2 miles per hour, or 220 feet per minute, and harnessed to a rope leading over a pulley and down a vertical shaft, could haul up a weight averaging 100 lbs., equaling 22,000 foot pounds per minute." The standard computation for horsepower–33,000 foot pounds, lifting one foot per minute–involves force, distance and time.[16]

To draw a vehicle, the horse must first propel its own weight. This exertion is then transmitted through the harness to the vehicle by the resistance of the collar or breastplate resting against the animal's shoulder. Attached to the breastplate or collar are traces, which are connected to the vehicle. As the horse moves forward, pushing its weight against the collar, this motion is transferred to the vehicle.

The weight of the carriage, the size and structure of the wheels, the distribution of the load and the friction of the surface over which the assemblage is drawn all influence the effectiveness of the draft. The less exertion required to draw a vehicle, the less the horse is impeded or restricted–or even injured. The governing concept was,

in the case of animal draught, that there are two principle objects of consideration; the first is, that of affording the moving power its greatest advantage, in giving motion to the carriage when at rest; which is done by placing the animal in a situation to effect a momentary exertion of his greatest muscular force: the second is the facilitating of the continuation of that motion when given, so that the constant exertion of the muscle force of the animal shall be as little as possible.[17]

Various experiments conducted during the horse-drawn vehicle era determined that undulating roads caused less strain on horses than those that were level, the theory being that the horse could alternate its efforts of exertion and experience intervals of rest, while the force of motion remained the same in regard to the carriage.[18]

Road surfaces were equally important to the effectiveness of actual horsepower. The cobblestones often used on city streets were not only hard, and thus concussive to the horse's feet and legs, but also slippery during inclement weather. City streets were generally detrimental to the horse's hooves, ligaments and tendons, precipitating numerous forms of lameness and diseases, such as navicular and ringbone.

The majority of the horses sold from a great market go to the city trade and are compelled to do their work on hard, unyielding pavements, pulling heavy loads, or developing speed that is an even greater strain on the feet and legs. The average period of usefulness on the city streets of a horse that was sound at the start is not more than five years, and it is manifestly decreased if he begins work in an unsound condition.[19]

Experiments with wooden plank roads proved unsuccessful, as the wood deteriorated and splintered under the continuous impact of hooves and wheels. Crushed stone roads also provided obstacles to the traction of wheels, since additional force was required to pull wheels over the stones. The "channel of conveyance," or surface over which a vehicle traveled, had such an influence on draft that it was generally included in its study, and the improvements in roads no doubt resulted largely from the desire to reduce resistance to horses in motion. On the other hand, wheeled vehicles were considered destructive to roads, because the wheels created deep ruts. Attempts to restrict the use of carriages were made and heavy taxes imposed from the seventeenth to the nineteenth centuries; both wide and narrow wheels were believed to be detrimental to roads, but according to the users of carriages, it was the inferiority of the roads that caused them to deteriorate. Attempts to improve road conditions responded to the persistent use of horse-drawn vehicles: "The first ones to agitate for good roads, and they who do most to see that good roads are provided, are the users of horses."[20]

The design of carriages also affected the horse's mobility. The elongated, unsprung and cumbersome vehicles used prior to about the middle of the eighteenth century were difficult for horses to draw. The improvements in vehicular design at that time resulted from attempts to lessen resistance and enhance the use of the horse's muscular power. In 1768, Richard Lovell Edgeworth "discovered, that springs of carriages were as advantageous to horses as to passengers."[21] This assessment was reiterated by Joseph Storrs Fry in his statement "The labor of the horses [is] lessened by the use of them [springs]."[22] Springs reduced the resistance upon the initial exertion to pull the vehicle, as well as providing comfort to the passengers. "With a spring interposed, a forward movement of the horse, more or less sudden, instead of being resisted by the entire weight of the carriage, is partially taken up by the spring and then gradually communicated to the vehicle."[23]

Some types of springs, however, increased the resistance for horses to pull against. The "C"-spring was considered to be particularly detrimental:

Carriages hung upon springs . . . which are called C springs, and which admit of very considerable longitudinal movement in the body of the carriage, are notoriously the most heavy to pull; and cabriolets, which are hung in this manner, are expressively called, in the stables, horse murderers, and require heavy, powerful horses to drag them, while lighter animals are able to drag much greater weights in stanhopes and spring-carts, which do not admit of this elasticity.[24]

Wheels were equally important in reducing the resistance in draft. As defined by Youatt, the wheel is a "strong circular frame of wood, composed of different segments, called felloes, [and] is bound together by a hoop, or several hoops of iron, called tires, which thus, at the same time

that it gives great strength, protects the outer surface from wear."[25] The introduction of wheels revolving around a fixed axle about 2500 B. C. in southern Mesopotamia represented an improvement over dragging or sliding vehicles such as sledges or the primitive travois. The spoked wheel, first used in early chariots, reduced the burden of draft previously imposed by heavy, solid wheels. Until the nineteenth century, however, the majority of wheels on European and American vehicles used for hauling or personal transportation were comprised of heavy, wide, partite members. The front wheels were usually of a reduced diameter, made to pass under the body or perch to facilitate turning. These small wheels required more rotations per foot to travel a given distance. The rear wheels were, by contrast, very high, thereby creating an unequal proportion. Even before the principle of weight reduction was applied to

Carriage Stuck in the Mud
1898
Floyd County, Indiana
Photograph Courtesy of State Historical Society of Wisconsin

The deplorable conditions of most roads concerned both carriage makers and carriage owners. This 1898 photograph of a carriage stuck in deep ruts on a muddy road in Floyd County, Indiana, was submitted to a contest sponsored by the League of American Wheel-men as part of the "Good Roads Movement." A prize was awarded to the image of the worst road condition, which would give visible evidence of the fact that an inferior road impeded both animal and vehicle, regardless of their quality.

the technology of wheel fabrication, it was acknowledged that "the wheels of carriages ought, for the ease of the horse as for the ease of the rider, to be made much lighter than those in common use."[26]

The development of the tire, which was introduced in various forms as early as the second millenium B.C. but was not commonly used until the eighteenth century, contributed to the reduction of the weight of the wheel; reinforcement allowed the nave, spokes and felloes to be made lighter without losing strength. The use of steam-bent felloes (especially those made of hickory, which possessed both strength and elasticity) instead of heavier partite members further decreased the weight of wheels. Machinery used in wheel fabrication after the industrialization of carriage manufacture allowed for precision in fashioning the parts of the wheels and contributed to weight reduction as well as decreased friction. Axles were also improved to reduce the exertion of horses while drawing a vehicle. Metal axles lubricated by oil to reduce friction were superior, it was claimed, to "clumsy wooden axles, smeared with a composition of tar and cow-dung, [in which] the friction may be so considerable as to neutralize a fifth, or fourth, or even more, of the strength of horses."[27] An advertisement for Synovial Axle Lubrication in *The Hub* reiterated the advantages of oiled axles with the motto "Save Your Horses."[28]

Reduction of the weight of all the elements of carriages was beneficial to horses, as lightness made vehicles easier to draw. Industrialized technology employed during the height of the Carriage Era (c. 1850-1910) responded to the recommendations of earlier advocates of lighter carriages who declared that "the weight of every carriage should be as little as possible [,] consistent with its requisite strength."[29]

As vehicles gradually became lighter, the improvements were acknowledged:

I am glad to see these Gothic vehicles, which are objects of terror to the traveler, and which are destructive in the highest degree to the roads, gradually giving way to light carriages, drawn by four horses, and driven, as coaches are, by a man upon the box. The owners have discovered the secret that horses in these carriages can draw

nearly as much weight, five or six miles per hour, as other horses, nearly twice as heavy, can draw in heavy waggons, about two miles per hour.[30]

The improvements in vehicles resulted not only from industrialized methods of fabrication but also from the study of the horse as an external source of energy. Design and technical elements, such as the parts of the running gear, were identified as either impediments or contributions to the exertions of the horse. The analyses of how the horse was utilized as a "medium for traction,"[31] recorded in various treatises on draft, influenced vehicular design.

Horses were combined in various ways to perform their role as the external source of energy for the vehicles they pulled. Of these, the single (one horse), pair (two horses abreast) and four-in-hand were most common, although tandem (one horse hitched in front of the other), unicorn (one horse hitched in front of a pair) and other combinations were used. The various ways in which horses were attached to a vehicle, as well as the specific size or breed, were intended to meet the requisites of the style of the vehicle and the efficiency with which it was mobilized.

The weight and size of a horse influenced its effectiveness in draft, as the load it could pull was proportional to its own weight and center of gravity. To this end, breeds or types were assigned to certain vehicles, as "the design and weight of the carriage should in large measure determine the size and build of the horse which is to draw it."[32] Like carriages, types of horses were classified as being appropriate for work or pleasure:

*Horses may, in general, be divided into two classes,–those kept for work, and those kept for pleasure. In the former class may be included farm-horses, stage, coach and omnibus horses, team-horses, employed in the transportation of goods, and moving heavy and bulky masses, carmen's horses,–and lastly the road horses of all professional men, who, like lawyers, doctors of medicine, and the like, are compelled to drive or ride many hour **per diem**, regularly, in the performance of their business. In the latter class may be included race-horses, match trotter, private gentlemen's saddle-horses, carriage horses, or roadsters, and many other animals belonging* to businessmen, which being employed during half the time or more in actual service, are used during spare hours on the road for purposes of amusement.[33]

Within the first class, specific draft breeds were established, such as the Percheron, Shire, Clydesdale, Belgian and Suffolk. Categorized within the pleasure class were horses such as the American standardbred, English hackney, and many other breeds of lightweight horses. Certain horses of a specific or mixed breed that demonstrated versatility could be considered suitable for many purposes:

> A very large proportion of the horses on the market are what are known in stockyard reports and market quotations as "general purpose" animals. They are horses without any particular type, embracing misfits of all kinds of breeding, though not scrubs, as a rule, and usually being of fair to good quality . . . . Such animals are used as the name designates, for anything and everything.[34]

The specific types or breeds of horses that evolved in different environmental conditions possessed particular characteristics in conformation, athletic potential and behavior. Domestication and continued selective breeding further enhanced features that could be utilized for specific functions. In this endeavor, types were developed for designated purposes; thus modification was especially significant to the relationship established between horses and carriages. Horse-drawn vehicles were designed for various functions that influenced their appearance, construction and seasonal use; similarly, breeds or types of horses were developed to conform to the vehicles they were intended to draw.

Hence it may be said that the division of horses into classes is the result of an endeavor to establish a balance of proportion between the horse and his work. With this end in view, various types of horses have been bred with the greatest care and attention to the development of those qualifications which render them particularly adaptable, in the combination of strength, symmetry, disposition and manners, for such specific work.[35]

Horses were frequently defined in reference to the type of vehicle they pulled; thus the

Proportions of the Thorough-bred and of the Cart Horse
c. 1880
From Modern Practical Farriery
Publisher, William MacKenzie, London, England; Glasgow and Edinburgh, Scotland

(following page)
Representative Types of Horses
1917
From New Standard Dictionary
Publisher, Funk & Wagnalls Company, New York and London

"The animal, whether for phaeton or plough, must be truly proportional and harmoniously constructed." *(Francis Ware, Driving)*

REPRESENTATIVE TYPES OF HORSES.

1. Pacing stallion.
2. Orloff stallion.
3. Thoroughbred stallion.
4. Percheron stallion.
5. Shire stallion.
6. French Coach stallion.
7. Shetland pony.
8. Hackney stallion.
9. Cleveland Bay stallion.
10. Belgian stallion.
11. Trotting stallion.
12. Arab stallion.
13. German Coach stallion.
14. Clydesdale stallion.
15. Suffolk stallion.

(For definitions of these varieties, see vocabulary.)

38

I

J

K

L

"brougham horse," "busser," "cart horse" or "coacher". A brougham horse

should be long and low, well-bodied, and from fifteen and a half to sixteen hands high, according to the height of the fore-wheels of the carriage; for it is important that the horse should match the carriage, as if he is too small he looks over-weighted (even if strong enough), and if too tall, the carriage looks out of proportion.[36]

The strictures regarding the type of horse employed to draw a vehicle were not always obeyed, however, for

every sort of horse may be seen in broughams: heavy brutes just fit for Pickford's vans; light weeds, more suited to a butcher's flying cart; prancing giraffes, that, if black, would be in place in a mourning coach; plodding cobs, travelling with necks poked out like a harnessed pig. Fortunately, many people are content with anything that will draw them, and no more think of looking at the form of a horse than at that of a locomotive steam-engine.[37]

In describing the omnibus horse, George Rommel wrote,

A typical busser stands from 15.1 to 15.3 hands, and weighs from 1,200 to 1,400 pounds. . . . In form he is a compact, rugged little horse with all the characteristics of the true draft type, set on short, strong legs, with ample bone, more or less feather, and sound feet. He should carry a fairly high head and show some style.[38]

The cart horse was stockier than the roadster and other types of carriage horses: "The beauty of the cart-horse depends not only on quality and symmetry, but on a sort of elephantine ponderosity that bespeaks power in every muscle and every limb."[39] Rommel describes a coacher as "considerably heavier, smoother, and more compact than a roadster. The conditions of his work, of course, require soundness as an absolute essential."[40]

Although not consistently adhered to, an ideal type appropriate for a given vehicle was understood in the process of selection:

Type-for-purpose has despite all difficulties, now, however, become a recognized grade in all the wholesale and retail marts of the country. If one enters a sale stable he asks for a "runabout"

Brougham and Pair: Winter Turnout
1892
From Moseman's Illustrated Guide for Purchasers of Horse Furnishing Goods
C. M. Moseman and Brother, New York, New York

horse, a "brougham" horse, a "phaeton" pair, and qualifies his requirements of height and shape in no way–the dealer knows just what he means.[41]

Each vehicle required an animal of a particular height, weight, strength and action, not only as an efficient means for drawing it, but also to harmonize with the overall appearance. The assemblage of horse, vehicle, harness, livery and servants, if appropriate to the style of the vehicle, constituted the turnout or the "correct appointment [which] may be defined as genuine harmony of detail and outline, quietness of colour and ornamentation, and appropriateness of animal, vehicle and equipment."[42]

The rigors of fashion and taste were not exclusive to elite turnouts; ideally all horses had to be in proportion to the vehicles they drew, in terms of both practicality and aesthetics.

The carriage is built to harmonise with the horses, and the horses are selected to set off the carriage well. The carriage and horses form a combination; and if they possess not harmonising parts, they are unsightly. A cart-horse in an elegant carriage is as much an anomaly as blood-horse in a heavy cart.[43]

Form was regarded as rational; good form was the objective of any turnout, from a simple cart to a full-liveried coach.

Once the pleasure drive, outing or journey was finished, the horse still had to be considered; the process that had made the horse a useful servant to man in turn increased its dependence on its keeper. Unlike an engine, a horse had to be fed and cared for whether or not it was used. To understand the difference between the horse and the automobile–aside from the most obvious contrasts between animal and machine such as speed (the horse averages about 8 mph at a trot, a car 55 mph at highway speed)–would require imagining the removal of the car engine from the automobile body when the vehicle was not in use. A horse-drawn vehicle could not be "turned off." Even at rest, the horse was capable of thought, discomfort, emotions, hunger and thirst. Furthermore, the interaction between humans and horses was symbiotic. The restrictions of domestication diminished the horse's ability to care for and

Burton Mansfield and His Tandem
c. 1900
Artist unknown
Watercolor on paper
Gift of Lida Bloodgood, 1957

Road Wagon and Trotters
1892
From Moseman's Illustrated Guide for Purchasers of Horse Furnishing Goods
C. M. Moseman and Brother, New York, New York

feed itself, and the uses to which it was assigned required artificial means to maintain its soundness, such as the application of horseshoes to protect its hooves from the impact of man-made roads. "The more man aims at this higher quality in his domestic animals, the greater the attention they must receive, so that man becomes the servant of his animals."[44]

In the nineteenth century horse care varied from indulgence to cruel neglect. It was recommended that the well-managed stable assign one man to care for three horses. Horses were groomed, dressed and bandaged, and the stalls were "set fair" with braided straw, colored sand and other elaborate decoration. Maintenance of a horse was a time-consuming, labor-intensive and often expensive task; in an article about horses published in *The New York Coach-Maker's Magazine*, Charles Dickens was quoted as saying, "No fine lady requires more constant waiting on than a horse. Other animals can make their own toilette; he must have a groom."[45] Feeding also had to occur at regular and scheduled intervals and in sufficient quantities. Ventilation, adequate light, bedding and blanketing were considered necessary amenities. By contrast, horses used to pull public conveyances such as streetcars were stabled below the ground floor or several stories above street level, in dark, poorly ventilated stables that were particularly vulnerable to fire.[46] These animals, who pulled overloaded vehicles, were often old, underfed and lame as a result of continuous work on hard, unyielding pavement. Although some horses–"slaves" or "screws"–who performed this sort of work were grade or inferior types, many were assigned to the job when their usefulness as carriage horses had subsided and they were relegated to increasingly difficult work.

The more active one goes into a country livery stable, where he is hacked about by people whose only interest in the beast is to take out of him the pound of flesh for which they have paid. He has no rest on week days, but his Sunday task is the hardest. On that sacred day, the reprobates of the village who have arrived at that perfect age of cruelty (which I take to be about nineteen or twenty) lash the old carriage horse from one public house to another, and bring him home exhausted and reeking with sweat. His mate goes into a job wagon perhaps, possibly into a herdic

**Stable of
H. H. Rogers, Esq.**
*1902
Architect,
Charles Brigham
From* The Architectural
Review:
Stables and Farm
Buildings

(opposite)
Fashion and Nature
*1886
From* Horse and Man
*Reverend J. G. Wood
Publisher,
J. B. Lippincott Compan
Philadelphia, Pennsylvan*

[cab], and is driven by night lest his staring ribs and painful lameness in his hind leg should attract the notice of meddlesome persons. The last stage of many a downhill equine career is found in the shafts of a fruit peddler's or junk dealer's wagon, in which situation there is continual exposure to heat and cold, to rain and snow, recompensed by the least possible amount of food.[47]

Finer carriage horses were also victims of cruelty; as indicators of social status and fashion, they were subjected to physical alteration. Ears were clipped in an attempt to make the head more refined. The flowing tail, which protects the horse from the irritation of flies, was "docked" (cut off at the tail bone) and "nicked" (permanently forced into a vertical position by severing the depressor muscles of the tail) to show off the hindquarters of the horse and prevent it from switching its tail over the reins. Abusive application of bearing reins and tight checkreins were introduced to the assemblage of harness to fix the horse's head in an unnatural position. Like the tight lacing endured by fashionable women of the nineteenth century, the misapplication of the checkrein was ridiculed and criticized as inflicting unnecessary pain, and "fashion" versus "nature" was a frequently cited conflict. In its natural movement, the horse uses its long neck muscle in the act of pulling; the checkrein, which connected to a bit in the horse's mouth, restricted neck and head motion and even a shake of the head could induce pain.

It is pitiful to go through the park or pass through the fashionable streets of our cities and see the sufferings which are endured by horses. While being driven round the park, stopping at fashionable stores and other places, horses may be seen undergoing this torture for hours together. The liveried brute and idiot of a coachman, of course, thinks it a very fine thing to sit behind these poor

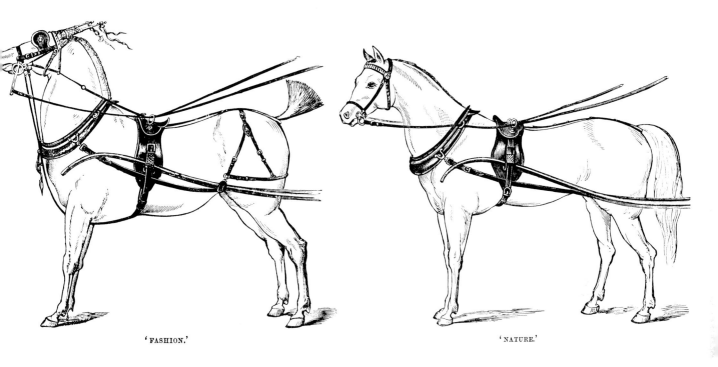

'FASHION.'   'NATURE.'

animals with their stuck up heads. I look at the arms of the carriage and I know who are the greatest fools of the upper class.[48]

In 1866 Henry Bergh introduced in New York the governing tenets of the American Society for the Prevention of Cruelty to Animals, which were based on those of the Royal Society for the Prevention of Cruelty to Animals, already established in England. Bergh set out to establish the civil rights of animals to "humane treatment" in the United States;[49] infractions witnessed by him were met with public reprimands and a notice to appear in court.

Enforcing sympathy and kind treatment of horses not only prevented the abuse of these creatures, but also was believed to improve morality and the level of civilization attained by humans. In addition to humane societies, another influencing factor was Charles Darwin's *On the Origin of Species by means of Natural Selection*, published in 1859, which linked human beings to their fellow creatures. Because animals had been considered either under the dominion of humans, to be used in whatever way the latter wished, or of such a low order that they did not feel pain, their sufferings had previously

been discounted or ignored. In addition, the horrible conditions endured by the poorer members of the human race in the nineteenth century in crowded slums, factories and mines hardened many against potential identification with the pain felt by animals: "Progress in humane feelings is hard to discuss during the decades when 25,000 streetcar horses died annually. Poverty and frustration in human society multiplied animal suffering."[50] Nevertheless, according to one observer, "The sooner all those engaged in the handling of live animals come to recognize that animals have a right to civil treatment, the sooner will they come to reap the full reward—both financial and moral—that comes to those who are entrusted with the care of, and association with, live animals."[51]

The prevalence of abusive treatment of horses as well as the labor intensity of their upkeep engendered hope that some alternate power would replace them. But in comparing the horse to the machine, which involved the cost of grain versus the cost of convertible fuels such as coal, it was determined that "animal power . . . is superior to any mechanical agent, and that beasts of

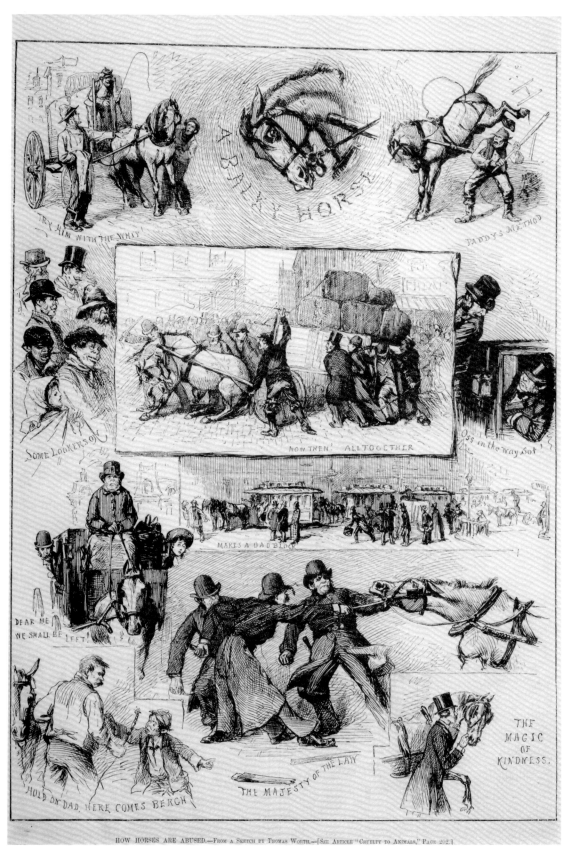

HOW HORSES ARE ABUSED.—FROM A SKETCH BY THOMAS WORTH.—[SEE ARTICLE "CRUELTY TO ANIMALS," PAGE 202.]

(left)

How Horses are Abused
March 27, 1880
Artist, Thomas Worth
From Harper's Weekly
Photograph Courtesy of State Historical Society of Wisconsin

The abuse of horses and Henry Bergh's influence a champion of animal right are graphically depicted in this illustration from a popular magazine.

(opposite)

Ridiculing the Bearing Rein
1894
From Facts for Horse Owners
D. Magner
The Magner Publishing Company, Battle Creek, Michigan

The bearing rein was an important piece of harnes that connected the bridle t the saddle, preventing the bridle from slipping off th horse's head. When tightly adjusted, it drew the horse's head into a painfu and exaggerated vertical position, a posture con- sidered fashionable by those who practiced "tight checking" and ridiculous cruel by those who oppose it. Bearing reins also prevented a horse from grazing and otherwise en- gaging its head. Advocate the bearing rein believed i forced the horse to throw weight into its hindquarte and thus produce the ext action associated with the popular "sensation horse or "high steppers" used i the show ring. Because th horse uses its long neck muscles in draft, howeve its pulling ability was impeded by a tight bearin rein.

draught, and particularly the horse, are not only the most ancient, but still remain the most advantageous source of power,"[52] and after centuries of associating with horses, it was difficult to imagine the possibility of establishing a relationship–let alone any measure of affection or pride–with a mechanical means of locomotion. Drawing an analogy between the steam engine and the horse, William Bridges Adams wrote:

Steam is a mere laborer–a drudge who performs his work without speech or sign, with clogged perseverance but without emotion. By dint of the garb in which he is clad, the machine which serves him for his body, he sometimes puts on the appearance of a live thing, shaking his polished metal clothing like an armoured knight; but this is only when he is stationary. His travelling garb is rough and rude, his breath is sulphureous, his voice is hissing, his joints creak, the anointing of his limbs gives forth an unpleasing gaseous odour; he carries with him a kitchen and a fuel chamber, and his whole appearance is black and unsightly. He may be personified when speaking of him; but no one pats his neck or speaks to him in a voice of

encouragement. It is not so with a horse or horses. They are beautiful and intelligent animals, powerful yet docile; creatures that respond to kindness, and shrink from cruelty or injustice.[53]

The development of the internal combustion engine by Carl Benz and Gottlieb Daimler in 1886, which would have a profound effect on the economy, society and ultimate decline of the horse-drawn transportation era, was not perceived initially as being a threat to the use of the horse as the motive power for vehicles:

Experiments with automobiles have shown that in their present state of development their effect on the horse market is not serious. For business purposes they have not yet been found to be entirely satisfactory . . . at present the "auto" is the least danger-ous of all the determining influences that operate against the horse market.[54]

It was also noted that

It is only in very recent times that the process of mechanical invention has begun to supersede some of the uses for which the strength and the speed of the horse for many thousands of years have alone

Drinking Fountain, New York City
March 26, 1881
From Frank Leslie's Illustrated Magazine
Photograph Courtesy of The New-York Historical Society, New York, New York

This fountain, donated to New York City in 1880 by philanthropist Olivia Eggleston Phelps Stokes, was designed to provide drinking water for both people and horses. Erected at the busy intersection of Madison Avenue and 23rd Street, it reflected the spirit of the Anglo-American animal protection movement. The fountain, rendered obsolete by the automobile, was dismantled in 1957 by New York City and subsequently removed to The Museums at Stony Brook where it stands, restored and functioning, as a monument to Miss Phelps Stokes's generous support for the humane treatment of animals.

been available. How far this commencing disestablishment of the horse from its unique position as the main agent by which man and his possessions have been carried and drawn all over the face of the earth will go, it is difficult to say at present.[55]

It was difficult to predict the supplanting of the horse by the automobile. Early automobiles were prohibitively expensive and perceived to be the novel, if not futuristic, toys of the very wealthy. The unfamiliar noise and odor associated with them made it "impossible to be truly dignified in even the most gorgeously appointed motor car."[56] But speed, power and a passion for the new increased their popularity, while improvements in manufacture and lowered prices began to increase distribution. By 1891, during the height of the horse-drawn Carriage Era, the following statement indicated the decline of the future of horses as a primary mode of transportation:

What is going to become of the horses? Of course everybody knows that horses were made to go, and from the way that steam, electricity, cables and other motive powers are increasing, it would seem that they were soon to fulfill their destiny.[57]

The PASSING of the HORSE

The silent horse power of this runabout is measurable, dependable and spontaneous— the horse-power generated by supplies of hay and oats is variable, uncertain and irresponsive. There is "*Nothing to watch but the road*" when you drive

The Oldsmobile

The Best Thing on Wheels

You see it everywhere. Doctors, Lawyers and Merchants find the Oldsmobile the most practical vehicle for business purposes. Ladies and children can readily understand its mechanism. Unvarying reliability proves it is built to run *and does it*

Price $650.00

Selling Agencies are established in all the larger cities where you will be gladly accorded the privilege of trying the Oldsmobile on the road. Write for illustrated book to Dept. Y.

Olds Motor Works
Detroit, Mich.
FACTORIES, DETROIT AND LANSING

NOTES

1 H. C. Merwin, *Road, Track and Stable* (Boston: Little, Brown, & Co., 1892) 178-179.

2 Horses have also been used in various capacities as riding or pack animals, or to pull vehicles that function as machines, such as farm implements, street scrapers and grinding apparatuses. Detailed and instructive publications on riding and driving, or the history of horsemanship, address more specific areas involving horses; this essay focuses in basic terms on the relationship between horses and vehicles, and thus omits information on the above, as well as more precise information on how to drive or harness the horse.

3 S. Bokonyi, L. Peel and D. E. Tribe, eds., *Domestication, Conservation and Use of Animal Resources* (Elsevier, 1983) 8. The author is indebted to Mary Aiken Littauer for providing this citation.

4 Edward L. Anderson and Price Collier *Riding and Driving* (New York: The Macmillan Co., 1905) 315.

5 Anderson and Collier 156-157.

6 William Henry Flower, *The Horse* (London: Kegan Paul, Trench, Trübner & Co., 1891) 39. The best contemporary source for the early evolution of the horse is George Gaylord Simpson, *Horses* (New York: Oxford University Press, 1951).

7 Richard Smythe, *The Mind of the Horse* (London: J. A. Allen & Co., 1967; 2d ed., 1972) 16.

8 William Youatt, *History, Treatment, and Diseases of the Horse* (Philadelphia: J. B. Lippincott Co., 1888) 410.

9 Merwin 1-2.

10 Francis Ware, *Driving* (New York: Doubleday, Page & Co., 1903) 120.

11 Youatt 405.

12 Henry William Herbert, *Hints to Horse-keepers* (New York: A. O. Moore & Co., 1859) 158.

13 Youatt 410.

14 William Bridges Adams, *English Pleasure Carriages* (London: Charles Knight & Co., 1837) 57.

15 J. Jacobs, *Observations on the Structure and Draught of Wheel-Carriages* (London: Printed for Edward and Charles Dilly, 1773) 2.

16 *Audel's New Mechanical Dictionary for Technical Trades* (New York: Theo. Audel & Co., n.d.) 349.

17 Jacobs 8.

18 Fairman Rogers, *A Manual of Coaching* (London: J. B. Lippincott Co., 1900) 186.

19 George Rommel, *Market Classes of Horses* (Washington: Government Printing Office, 1902) 16.

20 Anderson and Collier 165.

21 Richard Lovell Edgeworth, Esq., *An Essay on the Construction of Roads and Carriages* (London: J. Johnson & Co., 1813) 140.

22 Joseph Storrs Fry, *An Essay on the Construction of Wheel-Carriages* (London: Printed for J. and A. Arch, Cornhill; Baldwin and Co., Paternoster-Row, and Harding, St. James's Street; and T. J. Manchee, Bristol, 1820) 100.

23 Rogers 185-186.

24 Youatt 445.

25 Youatt 436.

26 Horatio Gates Spafford, A.M., *Some Cursory Observations on the Ordinary Construction of Wheel-Carriages . . .* (Albany: E. & E. Hosford, 1815) 1.

27 Fry 6.

28 *The Hub* 19:1 (1877).

29 Jacobs 10.

30 Fry 23.

31 Smythe 10.

32 James Garland, *The Private Stable* (Boston: Little, Brown & Co., 1892) 126.

33 Herbert 157.

34 Rommel 30.

35 Garland 123.

36 George Fleming, *The Practical Horse-Keeper* (New York: Cassell & Co., 1886) 13-14.

37 S. Sidney, *The Book of the Horse* (London, Paris & New York: Cassell, Petter, Galpin & Co., n.d., serial edition published 1875) 526.

38 Rommel 20.

39 Sidney 156.

40 Rommel 24.

41 Ware 247.

42 Ware 125.

43 Adams 199.

44 Wolf Herre, "The Science and History of Domestic Animals," *Science in Archaeology* (New York and Washington: Praeger Publishers, 1970) 259.

45 *The New York Coach-Maker's Magazine* 2.7 (1859): 157.

46 Gerald Carson, *Men, Beasts, and Gods: A History of Cruelty and Kindness to Animals* (New York: Charles Scribner's Sons, 1972) 92.

47 Merwin 21-22.

48 D. Magner, *Facts for Horse Owners . . .* (Battle Creek, Mich.: The Magner Publishing Company, 1894) 427-428.

49 Carson 99.

50 Carson 93.

51 Laurence M. Winters, *Animal Breeding* (New York: John Wiley & Sons, Inc., 2d ed., 1930) 9.

52 Youatt 409-410.

53 Adams 197-198.

54 Rommel 12.

55 Flower 2.

56 Ralph Straus, *Carriages and Coaches* (London: Martin Secker, 1912) 282.

57 *Varnish* (September 15, 1891) 404.